Trends & Developments in Selective Laser Melting (SLM) – A Detailed Review

J. Gunasekaran

Dr.P. Sevvel

Published by

BONFRING®
Intellectual Integrity

Trends & Developments in Selective Laser Melting (SLM) – A Detailed Review

9 788195 162659 >

ISBN 978-81-951626-5-9

Authors

J. Gunasekaran

Dr.P. Sevvel

Bonfring

309, 5th Street Extension, Gandhipuram,

Coimbatore-641 012.

Tamilnadu, India.

E-mail: info@bonfring.org

Website: www.bonfring.org

Message from Authors & Sincere Thanks

We kindly bring it to the Knowledge of the Readers that, this book entitled "Trends & Developments in Selective Laser Melting (SLM) – A Detailed Review" is a compilation of the various experimental research works carried out in various parts of the World by several researchers in the field of Selective Laser Melting (SLM). This book was compiled & framed by the Authors to provide a detailed knowledge about the various trends in the field of SLM. The Book contains a detailed review about the various trends, developments and experimental Research works carried out in the field of SLM. The Authors at any point of time, anywhere, does not claim that all these experimental works were carried out by them. The Authors of this book expresses their very Sincere Gratitude & Greater Thanks to All the experimental researchers whose experimental and investigational results are shared and mentioned in this book. Authors also register that, this Book was not compiled with the objective of earning Money or Profit or Fame in any manner, but purely, it was compiled with the objective of providing an awareness to the research community (especially to research scholars) about the trends and developments in the field of Selective Laser Melting (SLM). The Authors does not have any intention to misuse the experimental research findings of the researches (whose experimental results have been mentioned in this book) and the Authors also very kindly inform that, the details of websites, Names of each & every Researchers (whose experimental work have been described in this book) along with their experimental research works have been mentioned very clearly in the references section present at the end of each & every Chapter.

Dedicated to the Entire Scientific Research Community

and

to all Research Scholars

TABLE OF CONTENTS

LIST OF TABLES

LIST OF FIGURES

Chapter I

Introduction

CHAPTER I

1. Introduction

3D printing, or additive manufacturing, is the construction of a three-dimensional object from a CAD model. The term "3D printing" can refer to a variety of processes in which material is deposited, joined or solidified under control conditions using computers to create a three-dimensional object [1], with material being added together such as liquid molecules or powder grains being fused together, typically layer by layer to form products with the use of materials such as metal and its alloys, polymers and ceramics. 3D models have to be fetched to the machine in STL format.

In the 1980s, 3D printing techniques were considered suitable only for the production of functional or aesthetic prototypes, and a more appropriate term for it at the time was rapid prototyping. As of now, the precision, repeatability, and material range of 3D printing has increased to the point that some 3D printing processes are considered viable as an industrial-production technology, whereby the term additive manufacturing can be used synonymously with 3D printing. One of the key advantages of 3D printing is the ability to produce very complex shapes or geometries that would be otherwise impossible to construct by hand, including hollow parts or parts with internal truss structures to reduce weight.

In the early 2010s, the terms 3D printing and additive manufacturing progressed senses in which they were alternate shade terms for additive technologies, one being used in popular language by consumer-maker communities and the media, and the other used more formally by industrial end-use part producers, machine manufacturers, and global technical standards organizations. Until recently, the term 3D printing has been associated with machines low in price or in capability [2]. At present all over the globe, Additive Manufacturing (AM) is changing the way of organizations design and manufacture products. When used correctly, it can save impressive amounts of time and money.

1.1. History of AM

The history and development of rapid prototyping originate from two main preceding techniques, topography, and photo sculpture. Originating in the nineteenth century, these processes were initially very labour intensive. They later evolved to more modern techniques used in topography and metallurgy that resemble some current processes of rapid prototyping. These two techniques set the foundation for rapid prototyping as we know it today and their development as shown in table.1.1.

Table 1.1: History of AM

Year	Inventor	Process	Material	Company
1980s (patented in 1984)	Charles Chuck Hull	Stereo lithography (SLA)	liquid polymers	3D Systems corporation
1986	Deckard and Beaman	Selective Laser Sintering(SLS)	Powder polymers	3D Systems corporation
1988 (patented in 1992)	S.Scott Crump	Fused Deposition Modelling(FDM)	Plastics	Stratasys
1990	Researchers' at Massachusetts Institute of Technology(MIT)	Selective Laser Sintering(SLS) or Powder bed Process	Metal powders	Soligen Technologies, Extrude Hone Corporation
1995	Fraunhofer Society	Selective Laser Melting(SLM)	Metal powders	-

1.2. Transformation from Rapid Prototyping RP to AM

Earlier most of the technologies were made to make polymeric objects and had not been able to process other materials such as metals or ceramics. At later years, the development of AM technology based on metals and ceramics as discussed in table 1.2 are emerged which has been incorporated into industry and has made the global market.

Table 1.2: Transformation of Rapid Prototyping (RP) to AM

Year	Inventor	Process	Description	Company/Research
1994	Dr.Hans J.Langer	Direct Metal laser Sintering	First prototype using DMLS machine	Research work
1997	Sandia National Laboratories	Laser Engineered Net Shaping (LENS)	Commercialized	Optomec
2002	-	Electron Beam Melting(EBM)	Commercialized	Arcam AB

1.3. Additive Manufacturing Growth Rate

Since the emergence of these technologies and companies, the AM industry has been constantly expanding, growing, and advancing with much enthusiasm. With many industries seeing the lucrative value and abilities of AM, the market has been expanding very quickly. Many new types of RP and AM methods have been created since these original pioneers first started developing the technology. Some new technology has been novel and some just variations of the past types. There has also been a lot of development in the materials that can be used as well as research into making their properties optimal for end use.

These original technologies all started as RP, Layered Manufacturing (LM), or Solid Freeform Fabrication (SFF) methods, where they were designed to only be able to make quick prototypes using polymeric materials. Over the years, RP has moved into being AM where functional prototypes and parts can be made to perform in a variety of environments. The rise of AM being incorporated into industry has made it so the global market of AM has been on the rise since the first 3D printer was made. Figure 1.1 shows how the AM market has been increasing over the years. In 2012, the market was approximately $1850 million and is projected to almost double which reach $3475 million by 2017[3]. AM will become more integrated into industry and our personal lives as the technology and availability continue to grow.

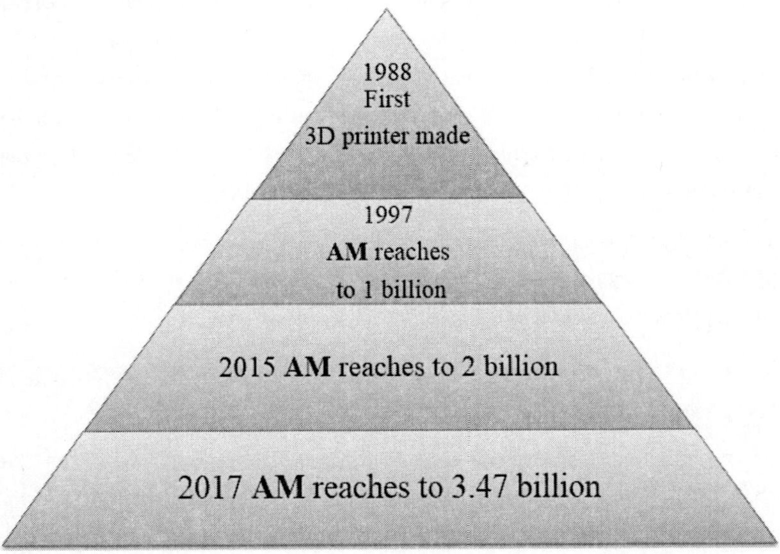

Fig. 1.1: Growth of Additive Manufacturing Industries [3]

1.4. Differences in Manufacturing Using Additive (AM) Subtractive Manufacturing (SM) and its Domination

AM technologies dominate SM in established manufacturing regions. Previous works, such as that of Connor et al [4] have determined the types of AM products that can fit into established manufacturing regions based on three contributing factors; customization, complexity advantage and volume. The three over-running manufacturing regions can be summarized as Mass Manufacturing, Manufacturing of the few and Customized for the Individual. However, there are key factors contributing the slow penetration of AM into the commercial market, rapid prototyping capability aside. These keys factors are namely the QA/QC methods or standards, and economic strategies.

1.4.1. Manufacturing Regions

Manufacturing regions can be categorically classified and defined by three key attributes; Complexity, Customization and Volume. Complexity is the final geometric complexity and feature location a manufacturing method can achieve. Customization is focused on the ease of feature and individualistic variability the manufacturing technology offers that makes a similar product unique to each other with a customizable feature [5]. Volume refers to the production quantity of parts in an order or batch, whereby the production volume can range from singular to multiple parts. Following its rapid advancement, the focus of AM has moved from rapid prototyping, to rapid tooling and manufacturing. Quite a few factors contribute to the selection of a method of manufacturing; cost, part complexity, material usage and material property requirements, time, energy consumption, sustainability. The Wohler's 2013 Report provides a widespread summary of AM's integration into the industrial market [6].

The study the degree of volume, complexity and customization has on an AM product vs. a traditionally manufactured product. Especially, where AM can dominate SM, or alternatively cannot compete with conventional methods.

1.4.2. High Production Volume based Manufacturing: Mass Manufacturing, Mass Complexity and Mass Customization

Low cost high production volume has been the primary focus of traditional manufacturing industries. The economic advantage of this region has enabled low-wage countries to optimize their manufacturing capability [7]. This is particularly true for Mass Manufacturing; parts characterized by their simplicity and lack of customizability. The high capital investment is required to create assembly and production lines using AM, does not make it a financially feasible investment for manufacturers. Conventional methods such as injection moulding still dominate this space. The complexity advantage offered by AM has provided opportunity to 'hit two mangoes with one stone' with the ability to eliminate forging, and joining process by printing complete parts in one operation. A case study of this is the metal ace tabular cup [8] used in hip replacement surgeries, where nearly 440,000 hip replacement surgeries are performed in the U.S. alone. The cup is one part of the implant inserted into the hip, previously manufactured using numerous SM and formative manufacturing methods such as forging, machining, coating etc. An AM powder-bed fusion metal process, can build the ace tabular cup, and add the porous surface into the surface layers in a single operation. Production of this product falls within the Mass Complexity category, where parts exhibit complex geometries, without customization features. The medical industry has capitalized on this manufacturing region for implants and prostheses through the replication of translated scans (x-ray, CT and MRI). The third high volume-based manufacturing region is Mass Customization, defined by its ability to provide individually designed products and services to every customer through high process agility, flexibility and integration. An agile and quick approach to changes is the key characteristic companies are after, in order to stay ahead of the market with innovation and competition [9]. The absence of lead time and quick design adaptation in AM, with unique characterization provided by 3D scanning, sets AM ahead of conventional manufacturing methods for this region. Popularly known products now produced through a mass customization manufacturing model.

1.4.3. Low Production Volume Manufacturing – Complexity for Free with Additive Manufacturing

In this Low-volume production of products, the product are manufactured by conventional methods will generally have limited complexity and customization; however with AM manufacturers can attain a high degree of complexity for the same cost. Low volume is defined specific to the industry, product and sale capacity [10]. The introduction of AM into the manufacturing sector began with low volume production for rapid prototyping, with Stereo lithography. Functional assembly manufacturing is still the primary use of AM according to the Wohler's report. However, investors are now interested in direct part manufacturing, particularly aerospace, automotive and rapid tooling industries. Hopkinson and Dickens et al [11], provide a successful cost model to illustrate the breakeven point of AM vs. SM for volume. The study examined the cost of fabricating a small plastic lever using AM powder bed fusion vs. conventional injection moulding manufacturing. While SM dominates the Mass manufacturing region financially, AM is better suited to fabricate tooling and fixtures required for conventional

Mass Manufacturing moulds. AM offers a reduced lead-time and cost to capitalize on high value, low production of parts such as those used in ships, automation, aviation, satellites etc. This concept is known as Rapid Tooling, referring to tools able to form several thousand or even millions of parts before wearing-out.

Direct and indirect AM technologies are used in case of rapid tooling. AM also offers manufacturers the ability to deal complexity for customizability in low production volume, where AM is being used for individually customized repair parts.

1.5. Additive Manufacturing and its Application Levels

The two application levels are

1. Rapid prototyping
   i. Solid Imaging or concept Modelling - Models
   ii. Functional prototyping – Checking and verifying
2. Rapid manufacturing

1.5.1. Rapid Prototyping

It is the application of AM (or) 3DPrinting to make prototypes and models or mock-ups, and rapid manufacturing is the application to make final parts and products. The manufacturing of tools, tool inserts, gauges, and so on usually is called rapid tooling.

If the final Part in a process is a direct product tends to direct manufacturing. If the final Part is not directly produced as a product, instead they are produced by an intermediate product say a pattern of a product which leads to direct tooling. Direct tooling Final parts are manufactured by means of die or mould said to be direct tooling. When usage of both type forms rapid tooling.

There are two sub categories

i. Direct tooling –direct tool
ii. Prototype tooling –creation of mould for e.g.

1.5.2. Rapid Manufacturing

Rapid Manufacturing is gaining a lot of momentum in the recent times. Today, customization plays a vital role which leads to rephrase this term today, people by mass customization. I would like to have a product which could satisfy my requirement. Future manufacturing is going to be more towards mass customization. When we talk about mass customization we cannot sacrifice time, we cannot sacrifice cost. So, we are still supposed to work to the expectation of the customer both in terms of economy and in terms of time delivery. So, in this scenario we are pushed from mass production to batch production to mass customised production. So, this mass customised production is termed or is redrafted and given a terminology called as Rapid Manufacturing.

We try to take the customer feedback or requirements converted it into a digital form; and that digital form is given directly to the machine the parts are produced and directly moved towards an application. We make a prototype for that prototype then we make tools and from those tools we try to go for the final product. So, Rapid Manufacturing is more towards mass customization where cost is not to be compromised, but things have to be produced to the customer's satisfaction with the highest integrity, that is Rapid Manufacturing.

Earlier Additive Manufacturing was thought of only from the prototyping point of view, today what has happened is the power source whatever you apply, the raw material whatever we use are now moving towards the real-time requirements, which was earlier thought about in plastic now it is changed into metal. Today the major challenge is shrinking product manufacturing time. The product life cycle time has shrunk down; the manufacturing time has shrunk down drastically.

1.6. Sustainable Manufacturing in Additive Manufacturing

Additive manufacturing enables eco-friendly products, it provides for end-of-product by giving solutions for eliminating waste and scrap and it allows for resource-efficient material choices via recycling. Future world should be a sustainable one so that we can live a sustainable life and for our future generations in terms of Manufacturing [12].

1.7. Steps Involved in Additive Manufacturing

AM/ 3Dprinting involves a number of steps that move from the virtual CAD description to the physical resultant part. Different products will involve AM in different ways and to different degrees. Small, relatively simple products may only make use of AM for visualization models, while larger, more complex products with greater engineering content may involve AM during numerous stages and iterations throughout the development process. Furthermore, early stages of the product development process may only require rough parts, with AM being used because of the speed at which they can be fabricated. At later stages of the process, parts may require careful cleaning and post-processing including sanding, surface preparation, and painting before they are used, with AM being useful here because of the complexity of form that can be created without having to consider tooling. Later on, we will investigate thoroughly the different stages of the AM process, but to summarize, most AM processes involve to some degree at least, the following steps as shown in Figure 1.2.

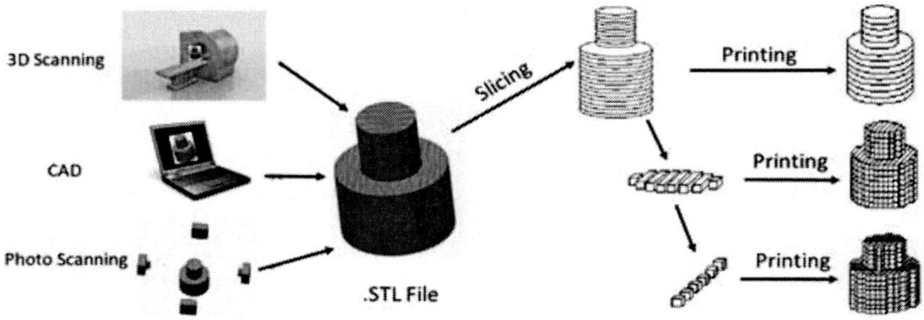

Fig. 1.2: General 3D Printing Process Steps [13]

1.7.1. *Conceptualization of Model and CAD*

All AM parts must start from a software model that fully describes the external geometry. This can involve the use of almost any professional CAD solid modelling software, but the output must be a 3D solid or surface representation.

Reverse engineering equipment (e.g., laser and optical scanning) can also be used to create this representation. The most common input method for AM technology is to accept a file converted into the STL file format originally built within a conventional 3D CAD system. There are however, other ways in which the STL files can be generated and other technologies that can be used in conjunction with AM technology. A few of these technologies are discussed here.

1.7.2. *Reverse Engineering (RE)Technology*

More and more models are being built from data generated using reverse engineering (RE) 3D imaging equipment and software.RE is the process of capturing geometric data from another object. These data are usually initially available in what is termed "point cloud" form, meaning an unconnected set of points representing the object surfaces. These points need to be connected together using RE software like Geomagic, which may also be used to combine point clouds from different scans and to perform other functions like hole-filling and smoothing[14]. In many cases, the data will not be entirely complete. Recently there have been huge improvements in scanning technology.

Engineered objects would normally be scanned using laser-scanning or touch probe technology. Objects that have complex internal features or anatomical models may make use of Computerized Tomography (CT)[15], which was initially developed for medical imaging but is also available for scanning industrially produced objects. This technique essentially works in a similar way to AM, by scanning layer by layer and using software to join these layers and identify the surface boundaries. Boundaries from adjacent layers are then connected together to form surfaces. The advantage of CT technology is that internal features can also be generated. High-energy X-rays are used in industrial technology to create high resolution images of around 1 μm. Another approach that can help digitize objects is the Capture Geometry Inside technology that also works very much like a reverse of AM technology, where 2D imaging is used to capture cross-sections of a part as it is machined away layer by layer. Obviously this is a destructive approach to geometry capture so it cannot be used for every type of product.

1.7.3. *Computer-Aided Engineering*

3D CAD is an extremely valuable resource for product design and development. One major benefit to using software-based design is the ability to implement change easily and cheaply. If we are able to keep the design primarily in a software format for a larger proportion of the product development cycle, we can ensure that any design changes are performed virtually on the software description rather than physically on the product itself. The more we know about how the product is going to perform before it is built, the more effective that product is going to be. This is also the most cost-effective way to deal with product development. If problems are only noticed after parts are physically manufactured, this can be very costly. 3D CAD can make use of AM to help visualize and perform basic tests on candidate designs prior to full-extent of commitment towards manufacturing. However, the more complex and performance related to design, the less likely we are to gain sufficient insight using these methods. However, 3D CAD is also commonly linked to other software packages, often using techniques like finite element method (FEM) to calculate the mechanical properties of a design, collectively known as Computer-Aided Engineering (CAE) software. Forces, dynamics, stresses, flow, and other properties can be calculated to determine whether the design performance under certain restrictions. While such software cannot easily predict the exact behaviour of a part, for analysis of critical parts a combination of CAE, backed up with AM-based experimental analysis, may be a useful solution. Further, with the advent of Direct Digital Manufacture, where AM can be used to directly produce final products, there is an increasing need for CAE tools to evaluate how these parts would perform prior to AM so that we can build these products right first time as a form of Design for Additive Manufacturing (DFAM).

1.7.4. *Haptic-Based CAD*

3D CAD systems are generally built on the principle that models are constructed from basic geometric shapes that are then combined in different ways to make more complex forms. This works very well for the engineered products we are familiar with, but may not be so effective for more unusual designs. Many consumer products are developed from ideas generated by artists and designers rather than engineers [16]. We also note that AM has provided a mechanism for greater freedom of expression. Problem we face today is that some computer-based design tools constrain or restrict the creative processes and that there is scope for a CAD system that provides greater freedom. Haptic-based CAD modelling systems [17], work in a similar way to the commercially available freeform modelling system to provide a design environment that is more intuitive than other standard CAD systems. An object can be seen on-screen, but also felt in 3D space using the Phantom software. The modelling environment includes what is known as Virtual Clay that deforms under force applied using the haptic cursor. This provides a mechanism for direct interaction with the modelling material, much like how a sculptor interacts with actual clay. The results using this system are generally much more organic and freeform surfaces that can be incorporated into product designs by using additional engineering CAD tools.

1.7.5. Conversion to STL, Transfer to AM Machine and STL File Manipulation

A critical stage in the additive manufacturing process that varies from traditional manufacturing methodology is the requirement to convert a CAD model into an (Stereo lithography) STL file. STL uses triangles or polygons to describe the object boundaries. There are several model limitations [18] that should be considered before converting a model to an STL file including physical size, water tightness and polygon count.

Once a STL file has been generated the file is imported into a slicer program. This program takes the STL file and converts it into G-code. G-code is a numerical control (NC) programming language. It is used in computer-aided manufacturing (CAM) to control automated machine tools (including CNC machines and 3D printers). The slicer program also allows the designer to customise the build parameters including support, layer height, and part orientation.

Nearly every AM machine accepts the STL file format, which has become a de facto standard, and nowadays nearly every CAD system can output such a file format. This file describes the external closed surfaces of the original CAD model and forms the basis for calculation of the slices.

The STL file describing the part must be transferred to the AM machine. Here, there may be some general manipulation of the file so that it is the correct size, position, and orientation for building.

There are several design considerations that must be evaluated when designing for additive manufacturing. These generally focus on feature geometry limitations and support or escape hole requirements and vary by technology.

1.7.6. Machine Setup

The AM machine must be properly set up prior to the build process. Such settings would relate to the build parameters like the material constraints, energy source, layer thickness, timings, etc.

1.7.7. Build

Building the part is mainly an automated process and the machine can largely carry on without supervision. Only superficial monitoring of the machine needs to take place at this time to ensure no errors have taken place like running out of material, power or software glitches, etc.

1.7.8. Removal

Once the AM machine has completed the build, the parts must be removed. This may require interaction with the machine, which may have safety interlocks to ensure that the operating temperatures are sufficiently low or that there are no actively moving parts.

1.7.9. Post-processing of Part or Product

Once the part or product removed and from the machine, It may require an amount of additional cleaning up before they are ready for use. Parts may be weak at this stage or they may have supporting features that must be removed. This therefore often requires time and careful, experienced manual manipulation.

1.8. Application

Parts may now be ready to be used. However, they may also require additional treatment before they are acceptable for use. For instance, they require priming and painting to give an acceptable surface texture and finish. Treatments may be laborious and lengthy if the finishing requirements are very demanding. They may also be required to be assembled together with other mechanical or electronic components to form a final model or product.

While the various stages in the AM process have now been discussed, it is important to realize that many AM machines require careful maintenance. Many AM machines use fragile laser or printer technology that must be carefully monitored and that should preferably not be used in a dirty or noisy environment.

Machines are generally designed to operate unattended, it is important to include regular checks in the maintenance schedule, and that different technologies require different levels of maintenance. It is also important to note that AM processes fall outside of most materials and process standards; explaining the recent interest in the ASTM F42 Technical Committee on Additive Manufacturing Technologies, which is working to address and overcome this problem [19]. However, many machine vendors recommend and provide test patterns that can be used periodically to confirm that the machines are operating within acceptable limits.

In addition to the machinery, materials may also require careful handling. The raw materials used in some AM processes have limited shelf-life and may also be required to be kept in conditions that prevent them from unwanted chemical reactions. Exposure to moisture, excess light, and other contaminants should also be avoided. Most processes use materials that can be reused for more than one build. However, it may be that reuse could degrade the properties if performed many times over, and therefore a procedure for maintaining consistent material quality through recycling should also be observed which tends to have a sustainable manufacturing. Figure 1.3 shows the various future application areas in industrial and service sectors.

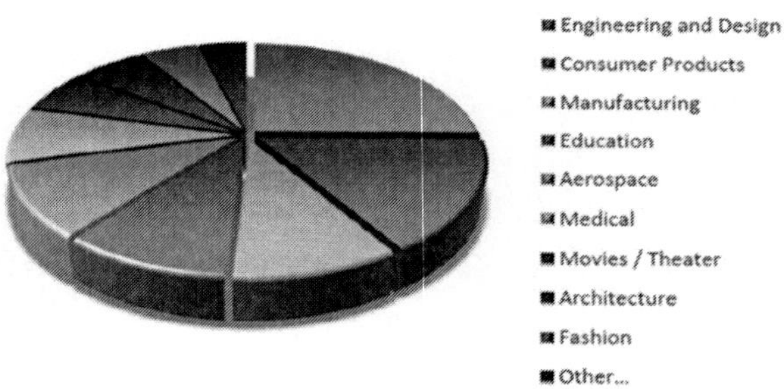

Fig. 1.3: Additive Manufacturing in various Industrial and Service Sectors [20]

1.8.1. AM in Aerospace & Defence

The aerospace and defence industry, the earliest adopters of 3D printing, with the first use of the technology going back to 1989. Now, three decades later, A&D represents a 16.8% share of the $10.4 billion additive manufacturing market and heavily contributes to on-going research efforts within the industry. An injector head is one of the core elements of propulsion modules, which forces the fuel mixture into the combustion chamber. Traditionally, injector heads are made from dozens or even hundreds of parts, which need to be machined and welded together. In contrast, AM printing enables these components to be manufactured as a single piece as shown in figure.1.4.

Fig. 1.4: Injector Head for Ariane 6 Launcher [21]

1.8.2. *AM in Medical Field*

The medical and dental industry is one among the growing adopters of AM. With 97% of medical AM professionals confident that the use of 3D printing will continue to increase within the sector, this trend seems set to continue. From medical devices to prosthetics and even bio printing, the applications of additive manufacturing for the medical industry are versatile and wide-ranging as shown in figure 1.5.

3D printing can be used to create custom prosthetic and orthopaedic devices from a number of certified biocompatible plastic or metal (e.g. titanium) materials. When it comes to implants, 3D printing is currently being used to create hip and knee joint replacements, cranial reconstruction implants and spinal implants. As of 2019, over 600,000 implants are estimated to be produced with the help of 3D printing. By 2027, this number is set to top 4 million [21].

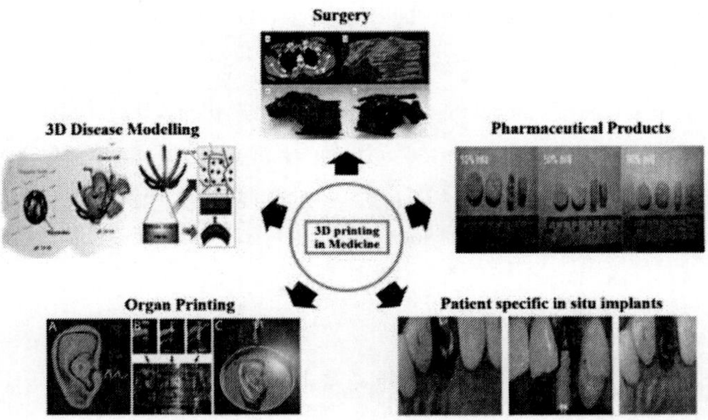

Fig. 1.5: Applications of AM in Implants, Prosthetics and Organ Printing [22]

1.8.3. AM in Automotive Field

Prototyping has become a key part of the product development process, offering a means to test and validate parts before they are manufactured. 3D printing offers a quick and cost-effective approach to designing and producing parts. Since the need for tooling is eliminated, product teams can significantly accelerate product development cycles with greater flexibility and customization. Figure 1.6 shows the bracket which has been fabricated using additive manufacturing in automotive application with features of reduced weight and increased strength which leads to sustainable product or parts.

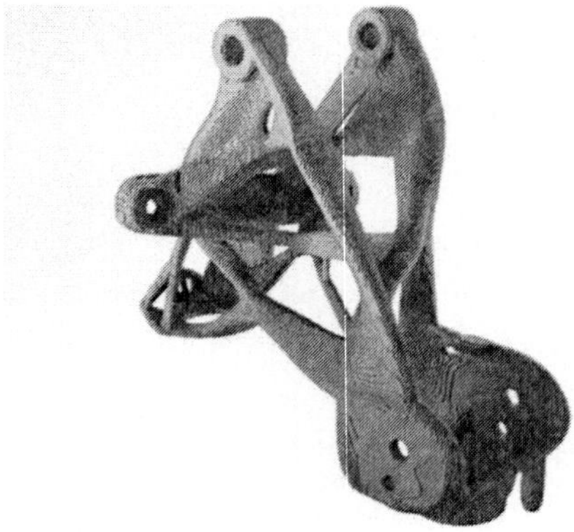

Fig. 1.6: Bracket Manufactured using AM Technologies [23]

1.9. General Factors in Design Considerations for 3D Printing

1.9.1. Overhangs

All 3D printing processes build parts layer-by-layer. Material cannot be deposited onto thin air, so every layer must be printed over some material.

Overhangs are areas of a model that are either partially supported by the layer below or not supported at all. There is a limit on the angle every printer can produce without the need of support material. For example, for FDM and SLA this angle is approximately 45° degrees [24]. It is a good practice to limit the overhangs of a model, as layers printed over support usually have a rougher surface finish.

1.9.2. Wall Thickness

The second thing to keep in mind when designing a part to be 3D printed is wall thickness. Every 3D printing process can produce accurately features that are thin up to a certain point. As a good practice, always add thickness to your models. Walls with thickness greater than 0.8 mm can be printed successfully with all processes.

1.9.3. Warping

Something that is often easily overlooked while designing a 3D model is the fact that the materials used for 3D printing undertake physical change: they are melted, sintered or scanned with a laser and solidified. The heating and cooling of material can cause the parts to warp while printing.

Large, flat surfaces can be especially prone to warping. Warping can typically be avoided by using correct machine calibration and having adequate surface adhesion between your part and the base bed. A good practice is to avoid large flat surfaces and add rounded corners to your 3D models. It is important to keep in mind what is the minimum feature size each 3D printing process can produce.

The minimum level of detail is connected to the capabilities and mechanics of each 3D printing process and to the selected layer height. The process and materials used will have an impact on the speed and cost of your product, so determining whether smaller details are critical to your model is an important design decision.

References

[1] Excell, john, *"The rise of additive manufacturing"*. The Engineer, 30 Oct 2013.

[2] "ISO/ASTM 52900:2015 – Additive manufacturing – General principles – Terminology".

[3] L. Wood, *"Global Additive Manufacturing Market* (2012–2017). Globe Newswire". Press Releases, CNBC LLC, Los Angeles, CA. 2014.

[4] Conner, B.P., et al., "Making sense of 3-D printing: Creating a map of additive manufacturing products and services". *Additive Manufacturing,* 2014. Vol. 1: pg. 64-76.

[5] Da Silveira, G., D. Borenstein, and F.S. Fogliatto, "Mass customization: Literature review and research directions". *International journal of production economics,* 2001. 72(1): pg. 1-13.

[6] Wohlers, T., Wohler's report 2013.

[7] Atzeni, E. and A. Salmi, "Economics of additive manufacturing for end-usable metal parts". *The International Journal of Advanced Manufacturing Technology,* 2012. 62(9-12): pg. 1147-1155.

[8] Unger, A.S., R.J. Lewis, and T. Gruen, "Evaluation of a porous tantalum uncemented acetabular cup in revision total hip arthroplasty: clinical and radiological results of 60 hips". *The Journal of arthroplasty,* 2005. 20(8): pg. 1002-1009.

[9] Da Silveira, G., D. Borenstein, and F.S. Fogliatto, "Mass customization: Literature review and research directions". *International journal of production economics,* 2001. 72(1): pg. 1-13.

[10] Additive Manufacturing Technologies: An Overview about 3D Printing Methods and Future Prospects. *Complexity,* 2019, pg.1–30

[11] Hopkinson, N. and P. Dicknes, "Analysis of rapid manufacturing—using layer manufacturing processes for production". *Proceedings of the Institution of Mechanical Engineers, Part C: Journal of Mechanical Engineering Science,* 2003. 217(1): pg. 31-39.

[12] I. Hager, A. Golonka, R. Putanowicz, "3D printing of buildings and building components as the future of sustainable construction" in *International Conference on Ecology and new Building materials and products, ICEBMP* 2016.

[13] Xu, Yuanyuan; Wu, Xiaoyue; Guo, Xiao; Kong, Bin; Zhang, Min; Qian, Xiang; Mi, Shengli; Sun, Wei , "The Boom in 3D-Printed Sensor Technology" , *Sensors,* Volume: 17 Issues: 5 1166, 2017 .

[14] Geomagic Reverse Engineering software. http://www.geomagic.com.

[15] Capture geometry inside. http://www.cgiinspection.com

[16] www.bathsheba.com

[17] Gao Z, Gibson, I. (2007). "A 6 DOF haptic interface and its applications in CAD". *Int J Comput Appl Technol.,* 30(3): 163–171.

[18] https://www.3dhubs.com/knowledge-base/3d-printing-geometry-estrictions/

[19] ASTM Committee F42 on Additive Manufacturing Technologies. http://www.astm.org/COMMITTEE/F42.html

[20] Steven Schain, "Real-World Applications of 3D Printing" Autodesk University.

[21] *Industrial Applications of 3D printing:* The Ultimate Guide.

[22] Roopavath, U.K., & Kalaskar, D.M. *"Introduction to 3D printing in medicine".* 3D Printing in Medicine, 2017, 1–20.

[23] Souvik das, *"3d printing electric car parts: the birth of an industry",* 2018 auto tech.

[24] https://www.3dhubs.com/knowledge-base/key-design-considerations-3d-printing/

Chapter II

Additive Manufacturing Techniques

CHAPTER II

2. Additive Manufacturing Techniques

Rapid prototyping (RP), one of the earlier additive manufacturing processes, allows for the creation of printed parts, not just models. A few of the noticeable advances this process offered to product development and production include the following: (1) reductions in time and costs, (2) enhanced human interaction, (3) the possibility of creating almost any shape that would otherwise be very difficult to machine, and (4) a shortened product development cycle. 47,48 It soon came to be referred to as direct fabrication of a part using a 3D model by way of additive manufacturing. As there are no tool setting, work piece fixture, and tool wear compensation requirements, the process was categorized as being the most flexible and direct form of digital manufacturing. In industry circles, the process began to be referred to as additive digital manufacturing (ADM) rather than rapid prototyping.

In the mid-1990s, the rapid prototyping processes became firmly established as indispensable components for the rapid development of a product. With the emergence and use of rapid prototyping, both scientists and students have been able to build and analyze models for the purpose of theoretical analysis, comprehension, and study. Medical doctors can build models of a diseased or injured human body for the purpose of analysis and determining the various test procedures to be used. With the aid of rapid prototyping, market researchers can receive quick feedback on what the ultimate end users think of a new product. The key steps involved in product development using rapid prototyping are shown in figure 2.1.

It is apparent in this figure that creating models more quickly can save a lot of time, and there are advantages to be gained by being able to test a number of models to eliminate flaws in their design and make any changes required for improved performance.

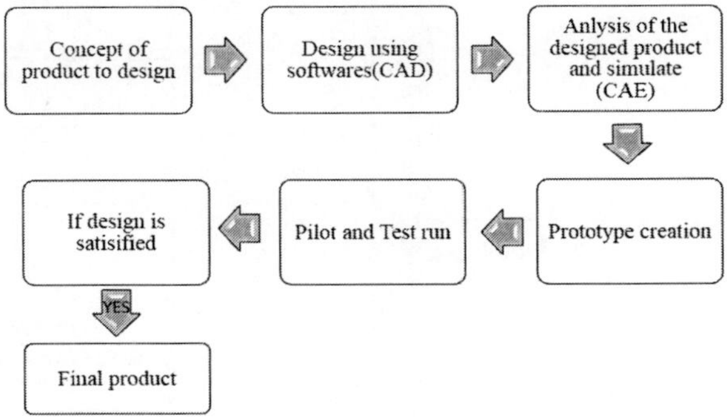

Figure 2.1: Representation of Life Cycle of Product

The technology of rapid prototyping has been used not just for the creation of models but also for the creation of finished products constructed of plastic materials. Subsequently, this technology has come to be termed 3D printing (3DP), although in reality it originated as rapid prototyping.

The advent of rapid prototyping and the concurrent emergence and use of the technologies of computer-aided design (CAD), computer numerical control (CNC), and computer-aided manufacturing (CAM) resulted in the evolution of rapid manufacturing (RM). The CAD, CNC, and CAM technologies, when combined together, made possible the printing of three-dimensional objects. Rapid prototyping was found not to be the best solution in all cases, especially when dimensions of the part are larger than the available additive manufacturing printers. This raised the need for the use of CNC machining processes. The materials often chosen and used for rapid prototyping are still very limited. It became clear following years of trials that it is possible to print parts made from both metals and ceramics but not from all of the available and commonly used manufacturing materials. Different additive manufacturing processes build and eventually consolidate the layers in different ways. A few of the processes depend on thermal energy from laser or electron beams that is directed via optics to melt or sinter (form a coherent mass without initiating melting) metal powder or plastic powder together. Figure 2.2 provides a pictorial overview of the various additive manufacturing processes.

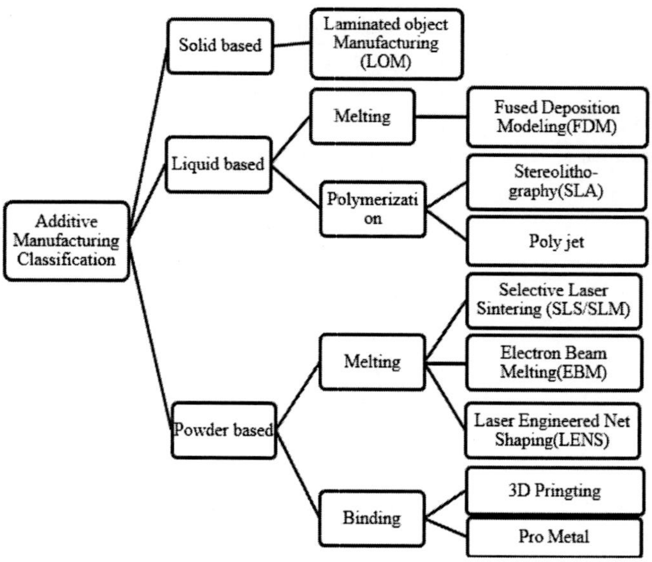

Figure 2.2: Various AM Process

In this figure, the processes are classified as liquid-based, solid-based, or powder based. The processes discussed in this chapter are considered to be most relevant and promising with regard to the overall future of this rapid development in technology for a broad spectrum of materials. Additive processes, which generate parts in a layered manner, first appeared in the 1980s with Stereolithography (SLA). Since then, several new ideas have been put forth, many patents have been approved, new processes have been invented, and a few have eventually been commercialized. An overview of the different processes is considered and briefly discussed in this section are

1. Stereolithography (SLA)
2. Fused deposition modelling (FDM)
3. Laminated object manufacturing (LOM)
4. Selective laser sintering (SLS)
5. Laser Engineered Net Shaping (LENS™)
6. Electron beam melting (EBM)

2.1. Stereo Lithography (SLA)

Stereo lithography is a 3D printing process which uses a computer-controlled moving laser beam, pre-programmed using CAM/CAD software. Stereo lithography (SLA) is an industrial 3D printing process used to create concept models, cosmetic - rapid prototypes, and complex parts with intricate geometries quickly [1, 2]. Parts can be produced in a wide selection of materials, extremely high feature resolutions, and quality surface finishes are possible with SLA. The material is also no longer limited to conventional polymers, but the fabrication of composites [3] and even metallic [4] or ceramic [5] specimens is also possible.

2.1.1. Principle

The basic principle governing the process of stereo lithography is photo polymerization. This is essentially a process where a liquid monomer or a polymer is converted into a solid by the application of ultraviolet light. The ultraviolet light acts as a catalyst for the reactions; the process is also known as ultraviolet curing. It has also been found to be applicable for powders of a ceramic suspended in a liquid.

2.1.2. Construction and Working

The SLA machine begins the 3D printing process by drawing the layers of the support structures, followed by the part itself, with an ultraviolet laser aimed onto the surface of a liquid thermoset resin. After a layer is formed on the surface of the resin, the building platform goes down, and a recoating bar moves across the platform to apply the next layer of resin. The process is repeated layer by layer until the build is complete as shown in figure 2.3.

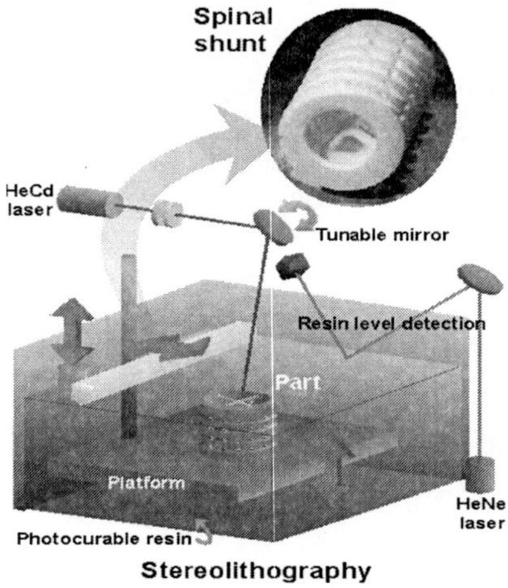

Figure 2.3: Typical SLA Process Machine [6]

Newly built parts are taken out to the lab where solvents are used to remove any additional resins. When the parts are completely clean, the support structures are manually removed. From there, parts undergo a UV-curing cycle to fully solidify the outer surface of the part. The final step in the SL process is the application of any custom or customer-specified finishing. Parts built in SL should be used with minimal UV and humidity exposure so they don't degrade.

Errors can be made during the process of Stereolithography. A significant error that often occurs during processing is over-curing, which tends to occur in overhanging parts because of the lack of fusion with a bottom layer. An error in the scanned line shape can be introduced during the scanning process. Because the resin used is a high-viscosity liquid, the thickness of the layer is often the only variable and errors can occur during position control. Other errors can occur when the part or component must have a specific surface finish that under normal circumstances would be provided manually.

When sustained research and development efforts eventually led to the possibility of using different materials to build a piece, the process became known as stereo lithography for multiple materials. In order to print with different materials, all of the resin has to be drained and new material provided when the process reaches the layer where the change must take place. This is essential even if the first material is to be used again as it is only possible to print consecutive layers.

When sustained research and development efforts eventually led to the possibility of using different materials to build a piece, the process became known as stereo lithography for multiple materials. In order to print with different materials, all of the resin has to be drained and new material provided when the process reaches the layer where the change must take place. This is essential even if the first material is to be used again as it is only possible to print consecutive layers.

2.1.3. *Free and Constrained Surface Approach*

In free surface approach, the building platform on which the printed part grows is located in a resin and coated with a liquid resin film. Illumination of the desired cross-section, which happens from above the resin bath, cures the first layer. After each layer, the platform with the growing part (i.e. the z-stage), is lowered further into the tank, and new resin is coated on top with a sweeper. This sets the stage for the subsequent layer [7].

The constrained surface approach or bottom-exposure approach has a building platform, which can be suspended above the resin bath. Illumination from below, through the transparent floor, cures a layer of resin between building platform and vat floor. This layer adheres to the platform, as the z-stage is raised by a defined distance. As each layer is cured, the building platform with the adhered part is elevated, and the part grows suspended from the platform downward [8]. As with the free surface setup, support structures made from the same printing material are needed in case of overhangs and undercuts to ensure adequate adhesion to the platform.

Recently, there has been a trend toward the bottom-exposure approach, as it has certain advantages [8, 9, and 10]. The smooth surface, which is created with a narrowly defined layer height due to the precise movement of the z-stage at an accuracy of down to 0.1–1 μm [11], is the main benefit in the constrained surface setup. Without the need for a sweeper, this layer can be created faster than in the free surface approach, reducing the printing time [9]. Another advantage, which decreases cost, is the lower amount of resin that is needed because the specimen does not have to be completely submerged in the vat.

A main disadvantage of the bottom-exposure setup, however, is that attractive forces between printed part and vat floor need to be overcome for each layer. When pulling up the z-stage, the newly cured layer needs to adhere to the layers above, and may not stick to the vat surface. Attempts to reduce this unwanted interaction include the application of hydrophobic coatings of the material tray [9, 11, and 12] and modification of the mechanical separation mechanism with tilting steps as well as the application of shear forces.

2.1.4. *Advantages*

- Competitively priced.
- Excellent surface finish.
- Easily duplicates complex geometries.
- One of the best surface finishes for an additive process.
- Material such as metals and ceramics can be used.

2.1.5. *Disadvantages*

- Photosensitive resin is a problem.
- Depending on the material, components may be brittle.
- Support structures can limit design freedom.
- Components are only UV-resistant to a limited extent.

2.1.6. *Applications*

SLA 3D printing accelerates innovation and supports businesses across a wide range of industries, including engineering, manufacturing, dentistry, healthcare i.e. bio medical [13], education, entertainment, jewellery, audiology, and more [14] are shown in figure 2.4 with the following as a requisite.

- Fit/form, proof of concept prototypes and engineering verifications.
- Investment Casting Patterns.
- Rapid Tooling, Jigs & Fixtures.
- Designer models, snap-fit assemblies.
- Scale & exhibition models.
- Optics, transparent covers.
- Molds & casting patterns.

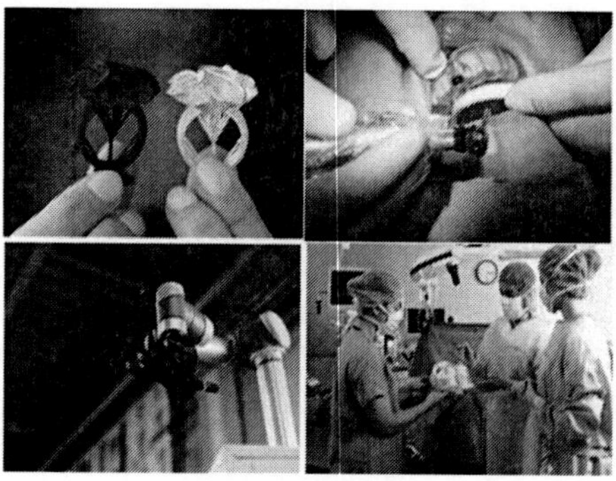

Figure 2.4: Application of SLA in various Fields [14]

2.1.7. *Comparison of various available Stereo Lithography Systems*

The various commercially available stereo lithography systems are indicated in the table 2.1. LCD-SLA is only available as a low-budget 3D printer for hobbyists. The more common methods such as conventional laser-SLA and DLP-SLA, however, have their low-cost and low-resolution desktop editions for consumers but can also offer high-end methods for professionals and industrial applications. While CLIP and TPP are both in higher price ranges, the former is designed for the extremely fast production of larger parts, while the latter is used for very slow fabrication of small parts with sub-micron resolution.

TPP - Two-photon polymerization (laser is used)

DLP - Digital light processing stereo lithography

CLIP - Continuous liquid interface production

LCD- Liquid crystal display stereo lithography

Table 2.1: Comparison of the Stereo Lithography System in Market [15]

SLA-type	Lateral resolution (μm)	Printing speed (mm/h)	Maximum print size (mm)	Commercial vendors
Laser	6–140	14	27–750	3DSystems, Formlabs, XYZ Printing,
TPP	0.400	—	100 × 100 × 3	Nano scribe
DLP	33–120	25–150	45–230	Envision TEC, Kudo3D
CLIP	50–100	500	80–320	Carbon
LCD	50–100	20–60	55–160	Spark Maker, Photo centric

2.2. Fused Deposition Modelling (FDM)

The patent for fused deposition modelling (FDM) (U.S. Patent 5121329) was awarded in 1992.It is an additive manufacturing technology commonly used for modelling, prototyping, and production applications. Fused filament printing is now the most popular process 3D printing [16,40]. Other techniques such as photo polymerisation and powder sintering may offer better results, but they are much more costly.

2.2.1. *Principle*

In this technique, liquid thermoplastic material is easily extruded from a movable FDM head and then deposited in ultra-thin layers onto a substrate [16]. Basically, the material is heated to one degree above its melting point so that it solidifies immediately following extrusion and also easily cold welds to the previous layers.

2.2.2. *Construction and Working*

FDM begins with a software process, developed by Stratasys, which processes an STL file (Stereolithography file format) in minutes, mathematically slicing and orienting the model for the build process. If required, support structures are automatically generated. The machine dispenses two materials – one for the model and one for a disposable support structure are created where needed, based on the part position and geometry as shown in figure 2.5.

The Fused Deposition Modelling (FDM) process constructs three-dimensional objects directly from 3D CAD data. A temperature-controlled head extrudes thermoplastic material layer by layer. This model is oriented and sliced into horizontal layers ranging from ± 0.127 - 0.254 mm thickness [17]. After reviewing the path data and generating the tool paths, the data is downloaded to the FDM machine. The system operates in X, Y and Z axes, drawing the model one layer at a time. This process is similar to how a hot glue gun extrudes melted beads of glue. The temperature-controlled extrusion head is fed with thermoplastic modelling material that is heated to a semi-liquid state. The head extrudes and directs the material with precision in ultrathin layers onto a fixtureless base. There is a wide variety of other AM process technology that can do this, including material extrusion, binder jetting, material jetting and directed energy deposition [18]. The result of the solidified material laminating to the preceding layer is a plastic model. Once the part is completed the support columns are removed and the surface is finished.

The exactly equivalent term, fused filament fabrication (FFF), was coined by the members of the Rep Rap project to give a phrase that would be legally unconstrained in its use [20]. This technology can also assist initiatives in sustainable development [21] since technologies are easily and economically [22] made from resources available to local communities.

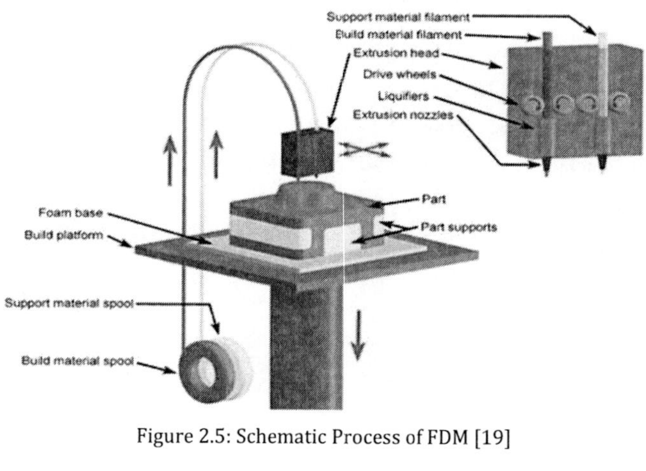

Figure 2.5: Schematic Process of FDM [19]

The exactly equivalent term, fused filament fabrication (FFF), was coined by the members of the Rep Rap project to give a phrase that would be legally unconstrained in its use [20]. This technology can also assist initiatives in sustainable development [21] since technologies are easily and economically [22] made from resources available to local communities.

Bound Metal Deposition (BMD) is an extrusion-based 3D printing technology developed by Desktop Metal, where metal components are constructed by extrusion of a powder-filled thermoplastic media. Bound metal rods—metal powder held together by wax and polymer binder—are heated and extruded onto the build plate, shaping a part layer-by-layer. Once printed, the binder is removed by debinding process, and then sintered—causing the metal particles to densify.

BDM technology most commonly uses metallic alloys such as stainless steels, tool steels, and other metals such as refractory metals, cemented carbides, and ceramics [23, 24].

2.2.3. Materials used for FDM

Several materials are available with different trade-offs between strength and temperature properties. As well as acrylonitrile butadiene styrene (ABS) polymer, polycarbonates (PC), polycaprolactone, polyphenylsulfones (PPSF), high density polyethylene (HDPE), acrylonitrile styrene acrylate (ASA), Polylactic acid (PLA), Glass and Carbon Fibre Reinforced Plastics (GFRP & CFRP) etc. as shown in table 2.2. A "water-soluble" material can be used for making temporary supports while manufacturing is in progress, this soluble support material is quickly dissolved with specialized mechanical agitation equipment utilizing a precisely heated sodium hydroxide solution.

Table 2.2: Materials used for Material Extrusion Technique (FDM) [25]

Material class	examples	Post-processing requirements
Thermoplastic polymers	ABS, HDPE, PPSF, PC, PTFE, PEEK, recycled plastics	Support removal
Polymer matrix composites	GFRP, CFRP	Support removal, curing
Ceramic slurries and clays	Alumina, Zirconia, Kaolin	Support removal, furnace drying and sintering
Green ceramic/binder mixture	Zirconia, Calcium phosphate	Support removal, debinding, sintering
Green metal/binder mixture	Stainless steel, Titanium, Inconel	Support removal, debinding, sintering
Biological materials	Bioink	-

2.2.4. *Advantages*

- No chemical post-processing required.
- Less expensive equipment due to no resins having to be cured.
- Use of materials resulting in an overall cost-effective process.

2.2.5. *Disadvantages*

- A seam line between the layers.
- They need for supports.
- Longer build-up time, requiring days to build complex parts.
- Delamination arising as a direct consequence of fluctuation in temperature.
- Low resolution of the Z-axis compared to other additive manufacturing techniques.

2.2.6. *Applications*

Compressor Variable Bleed Valve Duct Adapter, Ultem 9085 Air Duct used in aerospace [26] and prosthesis arm, manifold in carbon fibres for automotive[27] fields are produced as shown in figure 2.6 using FDM process.

Figure 2.6: Components Produced by FDM Process [25, 26]

2.2.7. *Commercially available FDM Machines*

The comparison of various FDM machines along its specification in table 2.3 which can be used for fabricating parts based on the customer requirements and the software's such as Grab cad print, Insight and control centre are used.

Table 2.3: FDM Machines with its Specifications [28]

Build Envelope	Stratasys F120	Stratasys F170	Stratasys F270	Stratasys F370	Stratasys F370
System Size/Weight	889 x 870 x 721 mm 124 kg	1626 x 864 x 711 mm 227 kg with consumables	1626 x 864 x 711 mm 227 kg with consumables	355 x 254 x 355 mm	355 x 305 x 305 mm
Part Accuracy	± .200 mm, or ± .002 mm/mm Whichever is greater			± .200 mm, or ± .002 mm/mm Whichever is greater	±0.127 mm or ± .0015 mm/mm whichever is greater

2.3. Laminated Object Manufacturing (LOM)

The patent for laminated object manufacturing (documented in U.S. Patent 4752352) was awarded in 1988. The latest developments of SL made possible the use of carbon fiber sheets and various composites, a technique mastered by the Envision TEC Company and the Impossible Objects start-up. The patent for laminated object manufacturing (documented in U.S. Patent US5730817A) was awarded in 1996 by Helisys Corporation [29].

These techniques are still being honed by their manufacturers and not yet widely spread.

2.3.1. Principle

The simplest LOM technique uses adhesive-coated sheet materials. Layers of material, plastic or layers of paper are fused, or laminated, together, with the application of heat and pressure. The material is coated with an adhesive layer and the feeding roller heats in order to melt the adhesive. The layer is then glued to the previous one to fabricate the part [30].

2.3.2. Construction and Working

The machine consists of four main components which are; placing of metal lamination, the heating system, the delivery and assembly station. The general lamination process starts with coated steel sheet layers are gradually added on a platform, then rectangular heater block apply pressure bonding purpose. The platform lowers by a depth equal to the sheet thickness of the first sheets and another sheet is advanced on top of the previously deposited layers.

The procedure repeated until the parts building is finished. The 3D parts are then manufactured by sequential lamination and the cutting of two-dimensional (2D) cross-sections using a laser beam as shown in figure 2.7. The velocity and focus of the laser beam are adjusted so the cutting depth corresponds to the thickness of the layer. By careful selection of both the velocity and focus, damage to the underlying layers is either minimized or avoided. To form an object, a heated roller is passed over the sheet of material on the build platform, melting its adhesive and pressing it onto the platform. A computer-controlled laser or blade then cuts the material into the desired pattern. The laser also slices up any excess material in a crosshatch pattern, making it easier to remove once the object is fully printed [31].

After one layer of the object is formed, the build platform is lowered by about one-sixteenth of an inch — the typical thickness of one layer. New material is then pulled across the platform and the heated roller again passes over the material, binding the new layer to the one beneath it. This process is repeated until the entire object has been formed.

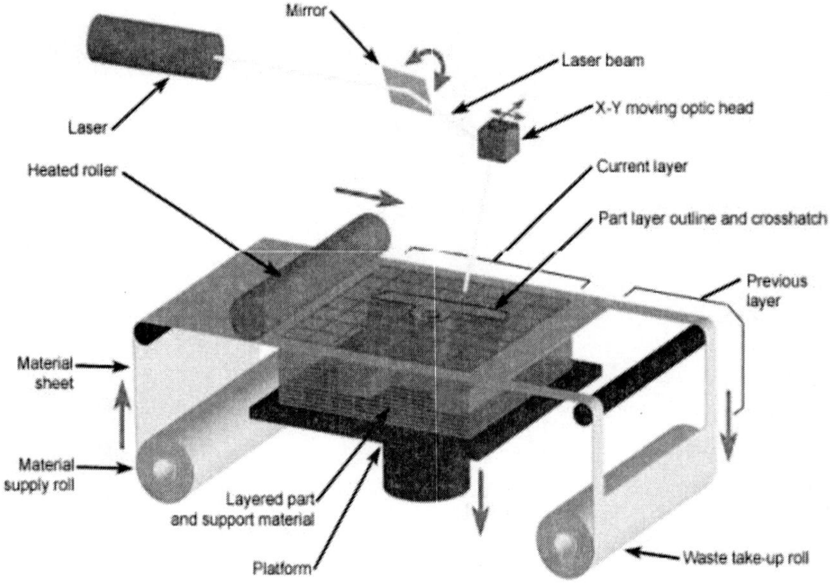

Figure 2.7: Schematic Layout of LOM Process [31]

Most importantly, this technique is relatively inexpensive, and toxic fumes are not generated. It offers the advantage of being automated, thus requiring minimal operator assistance. The density of the sintered ceramic part greatly depends on the lamination pressure and the porosity of the green ceramic tape, reaching up to 57% [32]. Hence a strong anisotropy in the sintered parts, resulting in a decrease of the flexural strength by up to 50% in lateral directions compared with the z-direction; prevail, although this renders the part susceptible to crack formation [33]. The primary limitation of this technique can be associated with accuracy problems along the Z-axis resulting in instability in dimensions.

This drawback limits the technical applications for LOM and is overcome by the development of cold pressure laminated object manufacturing, allowing for objects to be manufactured at room temperature with decreased pressures, avoiding inhomogeneous densification and rendering complex-shaped micro electron mechanical systems possible [34].

Further developments have also included investigations of LOM on curved surfaces [35], rendering this technology suitable for the economic production of a large scope of different applications.

2.3.3. Materials used for LOM

A variety of materials, including paper, ceramics, metals, plastics, fabrics, synthetic materials, and even composites, have been used.

2.3.4. Advantages

- Low cost.
- No need for post-processing or support structures.
- No deformation or phase change during the process.
- The possibility of building large parts.
- Environmentally friendly.
- Not health threatening.

2.3.5. Disadvantages

- Fabrication material is subtracted, thus causing wastage.
- The surface has low definition.
- Complex internal cavities are difficult to build.

2.3.6. Applications

LOM machines are used mainly for rapid prototyping plastic parts. Its low price and fastness makes it convenient to make prototypes, even though the produced objects are far from end-use parts. Mcor proposes a particular kind of LOM that they named Selective Deposition Lamination (SDL). It is a paper based technology that adds colour in the print as shown in figure 2.8.

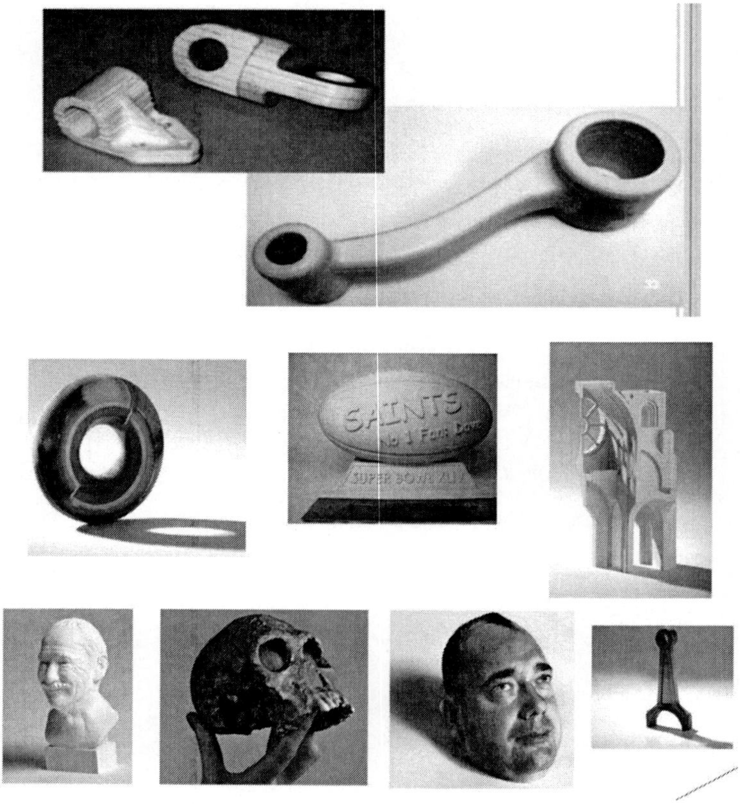

Figure 2.8: Products made by LOM Process [36]

2.4. Selective Laser Sintering (SLS)

Selective Laser Sintering (SLS) was developed at the University of Texas in Austin, by Carl Deckard and colleagues [37]. The technology was patented in 1989 and was originally sold by DTM Corporation. DTM was acquired by 3D Systems in 2001. The basic concept of SLS is similar to that of SLA.

2.4.1. Principle

An additive manufacturing layer technology, SLS involves the use of a high power laser to fuse small particles of plastic, metal, ceramic, or glass powders into a mass that has a desired three-dimensional shape. The laser selectively fuses powdered material by scanning cross-sections generated from a 3-D digital description of the part on the surface of a powder bed based on the design considerations [38]. After each cross-section is scanned, the powder bed is lowered by one layer thickness, a new layer of material is applied on top, and the process is repeated until the part is completed as per the requirement as shown in figure 2.9.

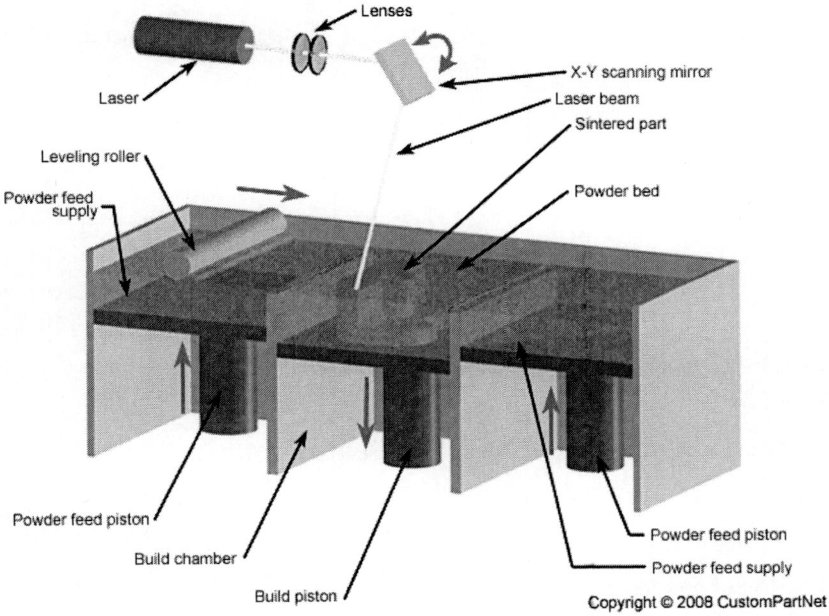

Figure 2.9: Schematic Layout of SLS Process [39]

2.4.2. Construction and Working

The Selective Laser Sintering (SLS) process uses a powdered polymer build material, typically nylon, is transferred from containers holding fresh powder material onto the build stage in the process chamber with a recoating tool. A high power laser then selectively scans the thin layer of powder, sintering together powder particles in the shape of the cross-section of the first layer of the 3D part. The build platform then descends one layer depth and the recoater transfers more fresh powder from the hopper to the surface of the first layer. Just like the first layer, the second cross-section of the 3D model is scanned and sintered [40]. The laser scanning process simultaneously generates the current layer and adjoins it to the previous layer, making a solid part. Because finished part density depends on peak laser power, rather than laser duration, a SLS machine typically uses a pulsed laser. The SLS machine preheats the bulk powder material in the powder bed somewhat below its melting point, to make it easier for the laser to raise the temperature of the selected regions the rest of the way to the melting point [41].

SLS does not need a separate feeder for support material because the part being constructed is surrounded by un-sintered powder at all times, this allows for the construction of previously impossible geometries. Also, since the machine's chamber is always filled with powder material the fabrication of multiple parts has a far lower impact on the overall difficulty and price of the design because through a technique known as 'Nesting' multiple parts can be positioned to fit within the boundaries of the machine. Further, in this process no post-curing is required, and the build time is noticeably fast.

2.4.3. *Materials for SLS*

A great variety of materials can be used. The materials normally chosen include (1) plastics, (2) metals, (3) combination of metals, (4) combination of metals and polymers, and (5) combination of metals and ceramics. Other materials that have been used are composite materials and reinforced polymers. For the specific case of metals, it is necessary to have a binder. Metal materials are not commonly used in SLS since the development of selective laser melting.

Currently available alloys used in the process include:

1. Stainless steel.
2. Maraging steel.
3. Cobalt chromium.
4. Nickel based alloy such as Inconel 625 and 718.
5. Aluminum alloy such as AlSi10Mg.
6. Titanium alloy such as Ti6Al4V.
7. Ceramics.

The quality of printed structures depends on the various factors include powder properties such as particle size and shape, density, roughness, and porosity [42]. The particle distribution and their thermal properties affect a lot on the flowability of the powder [43].

2.4.4. *Advantages*

- The sintered powder bed is fully self-supporting, allowing for:

  a. High overhanging angles (0° to 45 ° from the horizontal plane).

  b. Complex geometries embedded deep into parts, such as conformal cooling channels.

  c. Batch production of multiple parts produced in 3D arrays, a process called nesting.

- Parts possess high strength and stiffness.
- Good chemical resistance.
- Various finishing possibilities such as stove enameling, vibratory grinding tub coloring, bonding, powder, coating, flocking.
- Bio compatible according to EN ISO 10993-1 and USP/level VI/121 °C.
- Complex parts with interior components can be built without trapping the material inside and altering the surface from support removal.
- Fastest additive manufacturing process for printing functional, durable, prototypes or end user parts.
- Wide variety of materials with characteristics of strength, durability, and functionality.
- Due to the reliable mechanical properties, parts can often substitute typical injection molding plastics.

2.4.5. Disadvantages

- Parts have porous surfaces; these can be sealed by several different post-processing methods such as cyanoacrylate coatings, or by hot isotactic pressing.
- Only industrial SLS systems are currently widely available, so lead times are longer than other 3D printing technologies, such as FDM and SLA.
- SLS parts have a grainy surface finish and internal porosity that may require post processing, if a smooth surface or water tightness are required.
- Large flat surfaces and small holes cannot be printed accurately with SLS, as they are susceptible to warping and over sintering.

2.4.6. Applications

SLS technology is in wide use at many industries around the world due to its ability to easily make complex geometries with little to no added manufacturing effort. Its most common application is in prototype parts early in the design cycle such as for investment casting patterns, automotive hardware, and wind tunnel models. SLS is also increasingly being used in limited-run manufacturing to produce end-use parts for aerospace, military, medical, and electronics hardware. On a shop floor, SLS can be used for rapid manufacturing of tooling, jigs, and fixtures [44]. Because the process requires the use of a laser and other expensive, bulky equipment, it is not suited for personal or residential use; however, it has found applications in art.

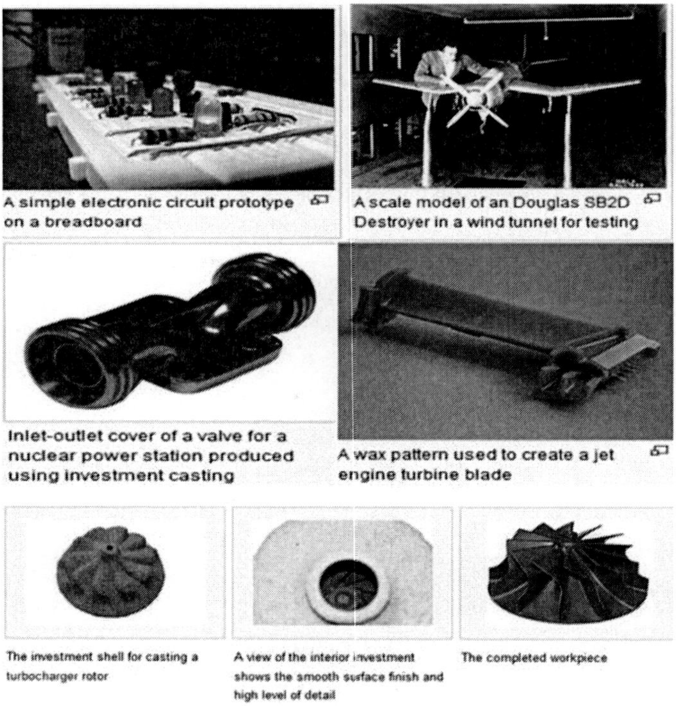

Figure 2.10: Prototypes, Patterns for Casting etc. Produced by SLS [45]

2.4.7. Commercially available SLS Machines

Table 2.4: Available SLS Machines with its Specifications [46]

Machines/ Specifications	**3D Systems sPro 230**	**EOS Formiga P110**	**EOS P396**	**EOSINT P760**
Build capacity	22 x 22 x 30 in / 550 x 550 x 750 mm	7.9 x 9.8 13in / 200 x 250 x 330mm	13.4 x 13.4 x 23,6 in / 340 x 340 x 600mm	27.6 x 15 x 22.9 in / 700 x 380 x 580mm
Build rate:	22 x 22 x 30 in / 550 x 550 x 750 mm	0.79 inches per hour / 20 mm per hour	up to 48 mm/h (1.9 in/h)	1.3 in per hour / 32mm per hour
Resolution/ Layer thickness	0.004 in / 0.1 mm	0024-.0047" / .06-.12mm	0.06 mm, 0.10 mm, 0.12 mm, 0.15 mm, 0.18 mm	.0024-.00709" / .06-.18mm
Feature highlight	Exceptionally fast printing speed.	Minimized energy consumption	EOS P 396 consumes for a typical building job (filling ratio 5.8%) **about 38% less energy and is up to 32% faster.**	Large build volume.

2.5. Laser Engineered Net Shaping (LENS)

As the most common type of Direct Energy Deposition (DED) system is a powder-based laser deposition system optimized for metals [47]. Laser Engineered Net Shaping (LENS) is a DED process was initially developed at the Sandia National Laboratories (SNL) in collaboration with Pratt & Whitney and was subsequently licensed to Optomec, Inc., in 1997. The patent for this technique (U.S. Patent 6046426) was awarded in 2000.

2.5.1. Principle

A laser beam created by a laser generator is focused through a lens onto the work piece. Metal powder, or metal wire, is introduced at the focus of the laser beam, where the metal powder and work piece melt and thus building on the surface. An inert gas is supplied to protect both the laser and the melt from contamination during the welding process. The cross section of the detail is thus built up step by step until the complete detail is created.

The method can also be used to repair damaged or worn surfaces and is then often described as laser welding with powder.

2.5.2. Construction and Working

A schematic of this process is shown in Figure 2.11. A high power laser is used to melt metal powder supplied coaxially to the focus of the laser beam through a deposition head. The laser beam typically travels through the centre of the head and is focused to a small spot by one or more lenses. The X-Y table is moved in raster fashion to fabricate each layer of the object. The head is moved up vertically after each layer is completed.

Metal powders are delivered and distributed around the circumference of the head either by gravity, or by using a pressurized carrier gas. An inert shroud gas is often used to shield the melt pool from atmospheric oxygen for better control of properties, and to promote layer to layer adhesion by providing better wetting on the surface [48]. Powder is delivered at the focal point of the laser into the controlled atmosphere chamber by an inert gas. The powder and laser create a melt pool on the surface of the substrate metal, which is then used for printing [49]. Because LENS is not a powder-bed technique; parts fabricated from this technique are not restricted to a powder-bed size and can be up to several feet in length. In Laser Engineered Net Shaping, a part is fabricated by focusing a high-powered laser beam onto a substrate with the primary objective of creating a molten pool into which metal powder particles are injected to gradually build each layer. The substrate is moved gently below the laser beam to deposit a thin cross-section and thereby create the desired part geometry.

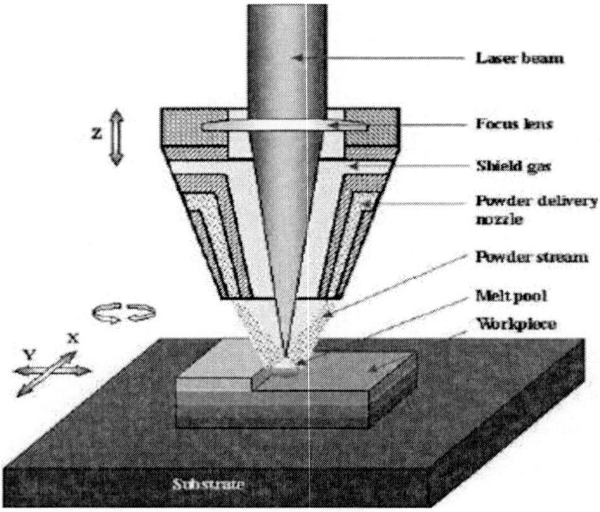

Figure 2.11: Schematic Layout of LENS Process [50]

Consecutive layers are then sequentially deposited to build a 3D part. By careful control of the

1. Fabrication parameters.
2. Desired geometric properties including both accuracy and surface finish.
3. Material properties, such as strength and ductility, a complex part can be easily built.

The LENS technique has been used to repair existing parts and structures, but it is primarily used to fabricate new ones. This technique does require a post-production process in that the finished part must be cut from the substrate.

2.5.3. Materials used for LENS

It can produce parts in a wide range of alloys, including titanium, stainless steel, aluminium, and other specialty materials; as well as composite and functionally graded materials [51, 52].

2.5.4. Advantages

- Less internal stresses.
- Ability to stack components in one job.
- Less support structures necessary than in laser-based systems.

2.5.5. Disadvantages

- Poor accuracy in growth direction.
- Poor surface finish.
- Cannot do very complex geometry.
- In need of more specialized process control for reliable deposits.

2.5.6. *Applications*

Primary applications for LENS technology include repair and overhaul, rapid prototyping, rapid manufacturing, and limited-run manufacturing for aerospace, defence, and medical markets [53, 54] with various benefits are as follows:

- Less heat input.
- Less distortion.
- Fewer propensities to crack.
- Excellent density & adhesion.
- Minimal effect on the component.

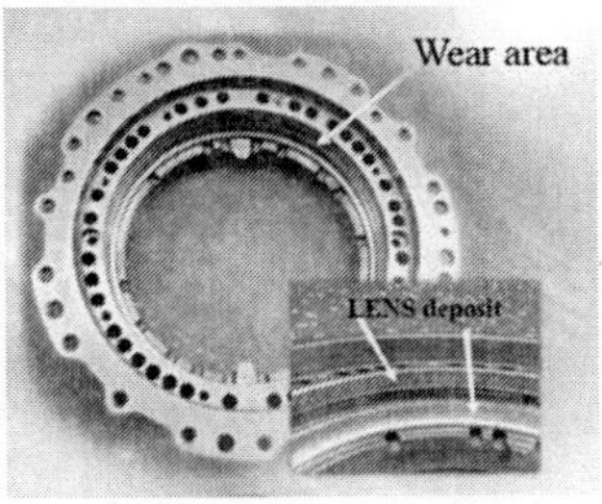

Low-wattage repair OF Ti-6Al-4V bearing housing

Repair of inconel 718 compressor or seal

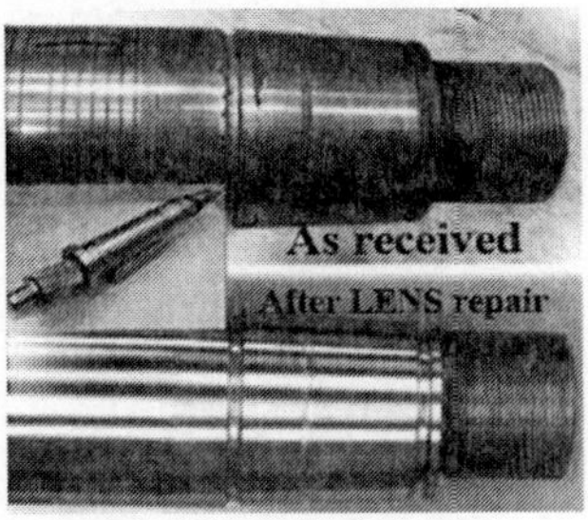

Comparison of thermal spray repaired shaft and LENS repaired shaft

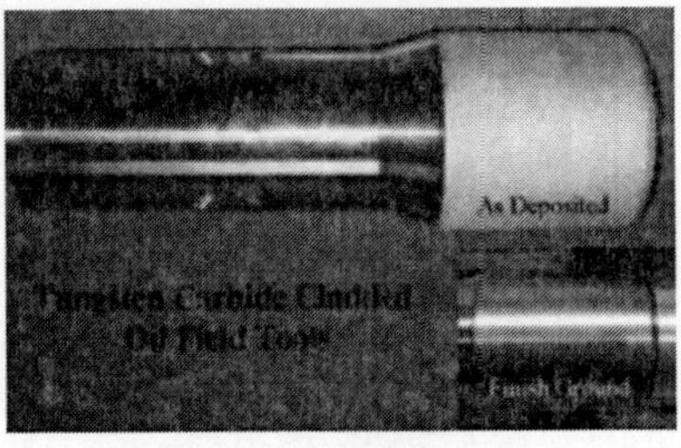

Oil field component clad with wear resistant tungsten carbide alloys And LENS

Figure 2.12: Areas where LENS Process is used for Producing Parts [53, 54]

2.5.7. Commercially available LENS Machines

With decades of experience in DED, Optomec incorporates this expertise into its LENS metal additive manufacturing solutions, resulting in a high-speed, high-quality, affordable metal 3D printing process making complex metal parts easier, more precise, efficient and affordable to produce and repair which are commercially available LENS systems are discussed in table 2.5.

Table 2.5: Commercially available LENS Systems [55].

Machines/ Specifications	**CS 150 AM CA**	**CS 800 AM CA**	**CS 600 AM CA**	**LENS 500 HYBRID CA SYSTEM**
XYZ Travel (mm)	150x150x150	800x600x600	600x400x400	350x325x500
Table Size XY (mm) / Payload (kg)	229x229 / 50	1300x650 / 150	1100x450 / 150	600x300 / 100
PosiBonal Accuracy (mm)	± 0.014	± 0.025	± 0.020	± 0.005
System Dimensions (mm)	2000x1000x1000	2997x2840x2662	2800x2700x2450	1650x2050x2050
CNC Controller	Win7/Galil	Siemens 840D	Siemens 840D	Siemens 828
Laser Power Standard (W)	400	1000 - 3000	1000 - 2000	500 -2000
Oxygen/Moisture Level (ppm)	<10	10	10	< 40

2.6. Electron Beam Melting (EBM)

Electron beam melting (EBM) is a metal additive manufacturing technology that uses an electron beam to melt layers of metal powder. First introduced in 1997 by Swedish company Arcam, EBM is ideal for manufacturing lightweight, durable and dense end parts [56]. The technology is primarily used within the aerospace, medical and defence industries.

2.6.1. Principle

Metal powders can be consolidated into a solid mass using an electron beam as the heat source. Parts are manufactured by melting metal powder, layer by layer, with an electron beam in a high vacuum. This powder bed method produces fully dense metal parts directly from metal powder with characteristics of the target material.

2.6.2. *Construction and Working*

Electron Beam Melting, like SLS and DMLS, belongs to the powder bed fusion family. However, in contrast to other metal AM technologies, which use a laser as their heat source[57], EBM uses a high-power electron beam to melt layers of metal powder. The melted layers of metal powder are then fused together to create a metal part as shown in figure 2.13.

A step-by-step view:

i. The build plate is coated with a layer of metal powder.
ii. As the layer is preheated, the powerful electron beam selectively melts powder in the areas defined by the digital CAD model.
iii. The next layer is then deposited and the beam melts and fuses layers together.

The process is repeated until the final shape of a part is achieved. After removing the excess powder, the metal part can then undergo post processing. To prevent contamination and oxidation of the powder, the printing process takes place in a vacuum environment.

2.6.3. *Materials for EBM Process*

A limited range of metals can be used with EBM, including with titanium alloys (ideal for medical implants), cobalt chrome, steel powders, and nickel alloy 718. These materials demonstrate high strength, resistance to corrosion and top mechanical properties, which is extremely valuable in stressful applications. It is important to note that any material used in EBM must be conductive, as the process heavily relies on electrical charges.

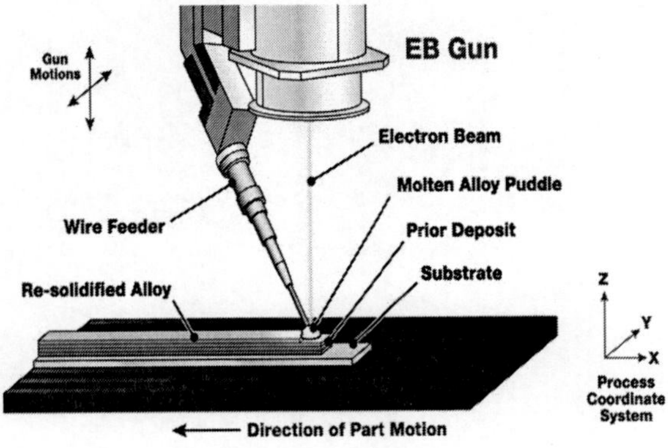

Figure 2.13: Schematic Layout of EBM Process [58]

EBM is also reportedly the only commercial AM solution for manufacturing titanium aluminide (TiAl) parts. TiAl is particularly noted for its light weight, strength and heat resistance, although it is prone to cracking [59, 60].

2.6.4. *Advantages*

- Manufacturing speed. The electron beam can separate to heat the powder in several places simultaneously, which significantly speeds up production. On the other hand, a laser must scan the surface point by point.
- Pre-heating the power before it melts limits the deformations and thus reduces the need for reinforcements and supports during manufacturing.

2.6.5. *Disadvantages*

- Precision. At the powder level, the electron beam is a little wider than the laser beam, which reduces the accuracy.
- The size of the parts that can be manufactured. Arcam's largest build volume (on the Q20 machine) represents a diameter of 350 mm for a height of 380 mm. On the other hand, laser machines (such as the X-Line of Concept Laser) offer manufacturing volumes at least twice as high.

2.6.6. *Applications*

EBM technology is mainly used in aeronautics and medical applications, particularly for implant design. Titanium alloys are particularly interesting because of their biocompatible properties and mechanical properties; they can offer lightness and strength. The technology is widely used to design turbine blades, for example, or engine parts as shown in figure 2.14. Electron Beam Melting technology will create parts faster than LPBF technology, but the process is less accurate and the finish will be of lower quality because the powder is more granular.

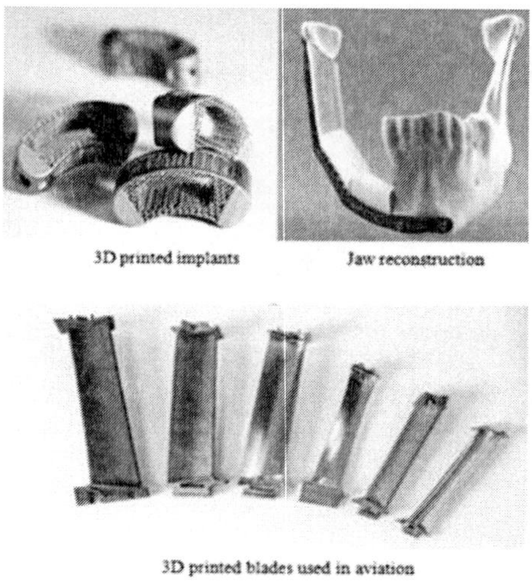

Figure 2.14: Application of EBM in Orthopaedic and Aerospace Fields [61]

2.6.7. *Commercially available EBM Machines*

EBM machines from Arcam EBM create dimensionally accurate parts quickly and efficiently by utilizing a high-power electron beam for high melting capacity and productivity. And the specifications are discussed in table 2.6.

Table 2.6: Commercially available EBM Machines from Arcam [62]

Machine/ Specifications	**Arcam EBM Spectra H**	**Arcam EBM Q10plus v2.0**	**Arcam EBM Q20plus**	**Arcam EBM A2X**
Max. build size	250 x 430 mm (D,H)	200 x 200 x 180 mm (W x D x H)	350 x 380 mm (Ø x H)	200 x 200 x 380 mm (W x D x H)
Max. beam power	6kW	3kW	3kW	3kW
Cathode type	Single crystalline	Single crystalline	Single crystalline	Tungsten filament
Min. beam diameter	-	140 μm	140 μm	250 μm
Max. EB translation speed	-	8,000 m/s	8,000 m/s	8,000 m/s
Active cooling	-	Water-cooled heat sink	Water-cooled heat sink	NO
Power supply	3 x 400 V, 32A, 13kVA	3 x 400 V, 32 A, 7 kW	3 x 400 V, 32 A, 7 kW	3 x 400 V, 32 A, 7 Kw
Size approx. (W x D x H)	1,328 x 2,344 x 2,858 mm	2,060 x 1,066 x 2,608 mm	2,400 x 1,300 x 2,945 mm	1,850 x 900 x 2,200 mm
Typical process temperature range	600-1,100°C	-	-	600-1,100°C

References

[1] Kodama H, "Automatic method for fabricating a three-dimensional plastic model with photo-hardening polymer". *The Review of Scientific Instruments,* 1981, 52(11), pg. 1770-1773.

[2] Hull CW, "Apparatus for production of three-dimensional objects by Stereolithography", 1984. US4575330B1.

[3] Wang X, Jiang M, Zhou Z, Gou J, Hui D. "3D printing of polymer matrix composites: A review and prospective. *Composites", Part B, Engineering.* 2017; 110, pg. 442-458.

[4] Bartolo PJ, Gaspar J, "Metal filled resin for Stereolithography metal part". *CIRP Annals. 2008,* 57(1), pg. 235-238.

[5] Halloran JW, "Ceramic Stereolithography: Additive manufacturing for ceramics by photo polymerization", *Annual Review of Materials Research,* 2016, 46(1), pg. 19-40.

[6] Jim H. Lee, *"Ceramic/polymer Composite Materials through Stereolithography, Ceramic Materials laboratory.*

[7] Han L-H, Mapili G, Chen S, Roy K, "Projection micro fabrication of three-dimensional scaffolds for tissue engineering", *Journal of Manufacturing Science and Engineering,* 2008, 130(2).

[8] Waheed S, Cabot JM, Macdonald NP, Lewis T, Guijt RM, Paull B, et al, "3D printed micro fluidic devices: Enablers and barriers", *Lab on a Chip,* 2016, 16(11): 1993-2013.

[9] Pan Y, Zhou C, Chen Y, "A fast mask projection Stereolithography process for fabricating digital models in minutes". *Journal of Manufacturing Science and Engineering,* 2012, 134(5): 51011.

[10] Huang Y-M, Kuriyama S, Jiang C-P, "Fundamental study and theoretical analysis in a constrained-surface stereolithography system", *International Journal of Advanced Manufacturing Technology,* 2004, 24(5-6), pg. 361-369.

[11] Hatzenbichler M, Geppert M, Seemann R, Stampfl J, "Additive manufacturing of photopolymers using the Texas Instruments DLP light crafter", In: Proceedings of SPIE, Emerging Digital Micro mirror Device Based Systems and Applications V. San Francisco: International Society for Optics and Photonics, 2013, pp. 86180A.

[12] Chen Y, Zhou C. Digital mask-image-projection-based additive manufacturing that applies shearing force to detach each added layer. 2013. US13872954

[13] Melchels, F.P.W., Feijen, J., & Grijpma, D.W, "A review on Stereolithography and its applications in biomedical engineering", *Biomaterials,* 2010, 31(24), 6121–6130.

[14] https://formlabs.com/blog/ultimate-guide-to-stereolithography-sla-3d-printing/#SLA%203D%20Printing%20Applications.

[15] Christina Schmidleithner and Deepak M. Kalaskar, *"Stereolithography",* 2018.

[16] "A Comprehensive List of All 3D Printing Technologies" MANUFACTUR3D. 2018-11-05.

[17] Gibson, I; Rosen, D.W; Stucker, B, "Additive Manufacturing Technologies: Rapid Prototyping to Direct Digital Manufacturing" 2010, Boston, MA: Springer.

[18] https://www.iitk.ac.in/dordold/index.php?option=com_content&view=category&layout=blog&id=200&Itemid=219

[19] Conner, B.P., Manogharan, G.P., Martof, A.N., Rodomsky, L.M., Rodomsky, C.M., Jordan, D.C., & Limperos, J.W, "Making sense of 3-D printing: Creating a map of additive manufacturing products and services". *Additive Manufacturing,* 2014, 1-4, pg. 64–76.

[20] http://www.kul3d.com/what-is-fdm-3d-printing/

[21] Jones, R., Haufe, P., Sells, E., Iravani, P., Olliver, V., Palmer, C., & Bowyer, A, "RepRap – the replicating rapid prototyper. *Robotica",* 2011, 29(01), pg. 177–191.

[22] Pearce, Joshua M.; et al, "3-D Printing of Open Source Appropriate Technologies for Self-Directed Sustainable Development". *Journal of Sustainable Development,* 2010, 3(4), pg. 17–29.

[23] Anzalone, G.C., Chenlong Zhang, Wijnen, B., Sanders, P.G., & Pearce, J.M, "A Low-Cost Open-Source Metal 3-D Printer", 2013, IEEE Access, 1, 803–810.

[24] Bose, A., Schuh, C.A., Tobia, J.C., Tuncer, N., Mykulowycz, N.M., Preston, A., Lund, A.C, "Traditional and additive manufacturing of a new Tungsten heavy alloy alternative". *International Journal of Refractory Metals and Hard Materials,* 2018, 73, pg. 22–28.

[25] https://en.wikipedia.org/wiki/Fused_filament_fabrication

[26] https://www.aero-mag.com/3d-printing-reaches-new-heights/

[27] https://diyconnect.net/popular-applications-of-fdm-3d-printing-technology/

[28] https://www.stratasys.com/3d-printers

[29] https://patents.google.com/patent/US5730817A/en

[30] https://thre3d.com/how-it-works/sheet-lamination/laminated-object- manufacturing-lom.

[31] Mekonnen, B.G., Bright, G., & Walker, A, "A Study on State of the Art Technology of Laminated Object Manufacturing (LOM)", *CAD/CAM, Robotics and Factories of the Future,* 2016, pg. 207–216.

[32] Y. Zhang, X. He, S. Du, J. Zhang, "Al2O3ceramics preparation by LOM (laminated object manufacturing)", *The International Journal of Advanced Manufacturing Technology* 17, 2001, pg. 531-534.

[33] M. Cho, A. Roosen, A. Stiegelschmitt, D. Schwanke, F. Bechtold, "In-situ shrinkage measurements of LTCC multilayers by means of an optical dilatometer", Key Engineering Materials I, *Euro Ceramics VII,* 2002, pg. 1281-1284.

[34] K. Schindler, A. Roosen, "Manufacture of 3D structures by cold low pressure lamination of ceramic green tapes", *Journal of the European Ceramic Society,* 2009, 29 (5), pg. 899-904.

[35] D. Klosterman, R. Chartoff, N. Osborne, G. Graves, A. Lightman, G. Han, A. Bezeredi,S. Rodrigues, "Development of a curved layer LOM process for monolithic ceramics and ceramic matrix composites", *Rapid Prototyping Journal* 5(2), 1999, pg. 61-71.

[36] http://www.mkstechgroup.com/laminated-object-manufacturinglom/

[37] Deckard, C., "Method and apparatus for producing parts by selective sintering", U.S. Patent 4,863,538

[38] "Design Guide: Selective Laser Sintering (SLS)" Xometry.

[39] Anoop Verma, and Rahul Rai, "Energy Efficient Modelling and Optimization of Additive Manufacturing Processes", solid freeform fabrication symposium, Austin, TX, 2013, pg.231-241.

[40] Kruth, JP.; Mercelis, P.; Van Vaerenbergh, J.; Froyen, L.; Rombouts, M, "Binding mechanisms in selective laser sintering and selective laser melting", *Rapid Prototyping Journal,* 2005, 11(1): pg. 26–36.

[41] S. Narayanan, "GCMM 2004: 1st International Conference on Manufacturing and Management", 2005.

[42] Leturia, M.; Benali, M.; Lagarde, S.; Ronga, I.; Saleh, K, "Characterization of flow properties of cohesive powders: A comparative study of traditional and new testing methods", *Powder Technology,* 2014, Vol 253: pg. 406–423.

[43] Leu, M.C., Pattnaik, S., & Hilmas, G.E, "Investigation of laser sintering for freeform fabrication of zirconium diboride parts", *Virtual and Physical Prototyping,* 2012, 7(1), pg. 25–36.

[44] "Selective Laser Sintering Applications Overview |Quickparts". www.3dsystems.com

[45] https://en.wikipedia.org/wiki/Selective_laser_sintering#Applications3

[46] https://www.sculpteo.com/blog/2017/03/14/list-of-professional-3d-printers/

[47] R.P. Mudge and N.R. Wald, "Laser Engineered Net Shaping Advances Additive Manufacturing and Repair", Weld. J.N.Y., 2007, 86(1), p 44

[48] J.E. Shugeresky, D.M. Keicher, J.A. Romero, Griffith, L.D. Harwell, "Using Laser Engineering Net Shaping(LENS) process to produce complex components from a CAD Solid model", SPIE Proceedings, 1997, *Laser as tools for Manufacturing II,* Vol. 2993.

[49] J. Irwin, E.W. Reutzel, P. Michaleris, J. Keist, and A.R. Nassar, "Predicting Microstructure from Thermal History during Additive Manufacturing for Ti-6Al-4V", *Journal of Manufacturing Science and Engineering,* 2016.

[50] Chang, K.-H, *"Rapid Prototyping. e-Design",* 2015, pg. 743–786.

[51] Onuike, B., Heer, B., & Bandyopadhyay, A, "Additive manufacturing of Inconel 718—Copper alloy bimetallic structure using laser engineered net shaping (LENS)". *Additive Manufacturing,* 2018, vol.21, pg. 133–140.

[52] Zhang, Y., Wu, L., Guo, X., Kane, S., Deng, Y., Jung, Y.-G. Zhang, J, "Additive Manufacturing of Metallic Materials: A Review. *Journal of Materials Engineering and Performance",* 2017, Vol. 27(1), pg. 1–13.

[53] dhttps://www.rpmandassociates.com/uploads/lens_advances_manufacturing_and_repair_.pdf.

[54] R.R. Mudge and N.R. Wald, "Laser engineered net shaping advances additive manufacturing and repair", 2007, *Welding Journal* 86 (1), pg. 44-48.

[55] https://optomec.com/3d-printed-metals/lens-printers/

[56] L.E. Murr, S.M. Gaytan, D.A. Ramirez, E. Martinez, J. Hernandez, K.N. Amato, P.W. Shindo, F.R. Medina, and R.B. Wicker, "Metal Fabrication by Additive Manufacturing Using Laser and Electron Beam Melting Technologies", *J. Mater. Sci. Technol.,* 2012, 28(1), p 1–14.

[57] M. Jamshidinia, F. Kong, and R. Kovacevic, "Numerical Modeling of Heat Distribution in the Electron Beam Melting_ of Ti-6Al-4V", *J. Manuf. Sci. Eng.,* 2013, 135(6), p 061010

[58] http://additivemanufacturing.com/2015/10/14/electron-beam-additive-manufacturing-ebam-advantages-of-wire-am-vs-powder-am/.

[59] Zhang, L.-C., Liu, Y., Li, S., & Hao, Y, "Additive Manufacturing of Titanium Alloys by Electron Beam Melting: A Review". *Advanced Engineering Materials,* 2017, 20(5), 1700842.

[60] Lewandowski, J.J., & Seifi, M, "Metal Additive Manufacturing: A Review of Mechanical Properties", *Annual Review of Materials Research,* 2016, 46(1), pg. 151–186.

[61] https://www.ge.com/additive/ebm

[62] https://www.ge.com/additive/additive-manufacturing/machines/ebm-machines

Chapter III

Selective Laser Melting

CHAPTER III

3. Selective Laser Melting (SLM)

Selective laser melting, also known as direct metal laser melting/Sintering (DMLM/DMLS) or laser powder bed fusion (LPBF) is one of the printing technologies started in 1995 at the Fraunhofer Institute ILT in Aachen, Germany, with a German research project, resulting in the so-called basic ILT SLM patent DE 19649865 for metallic materials. Already during its revolutionary period Dr. Dieter Schwarze and Dr. Matthias Fockele from F&S Stereolithographietechnik GmbH collaborated with the ILT researchers Dr. Wilhelm Meiners and Dr. Konrad Wissenbach.

3.1. Introduction

Selective Laser Sintering and Direct Metal Laser Sintering are same terms, with SLS used to refer to the process as applied to a variety of materials such as plastics, glass, ceramics, whereas DMLS refers to the process as applied to metal alloys. But what sets sintering apart from melting or 'Cusing' is that the sintering processes do not fully melt the powder, but heat it to the point that the powder can fuse together on a molecular level. And with sintering, the porosity of the material can be controlled.

Selective Laser Melting, on the other hand, can do the same as sintering and go one step ahead, by using the laser to achieve a full melt. Meaning the powder is not merely fused together, but is actually melted into a homogeneous part. That makes melting the way to go for a monomaterial, as there's just one melting point, not the variety you'd find in an alloy. If you're working with an alloy of some sort, you will go for SLS or DMLS or if you are working with, pure metal, you can go for SLM.

3.2. Principle

The SLM process has the ability to fully melt and fuse the metal into a solid three-dimensional part by using a laser source based on the STL file which is fed into the system.

3.3. Working Process

It (SLM) is a comparatively different technology that is used to form metal parts through the use of applying a laser beam to a powder bed. This allows for various material crystalline structure arrangements to be possible that might otherwise be impossible with conventional metal removing processes. This stimulates a higher density part with better-quality strength and durability. Due to complete melting of produced parts typically have better surface evenness and more regular microstructure arrangements. It is a powder bed fusion process, which routines high intensity laser as an energy source to melt and fuse selective regions of powder one over the other, rendering to the design (CAD) data[1, 2].

The SLM process consists of a sequence of steps from CAD data preparation to removal of fabricated component from the platform. Before the CAD data are uploaded to the SLM machine for production of components. The STereo Lithography (STL) files have to be processed by software, to provide support structures for any overhanging features and to generate slicing data (codes for corresponding design) for laser scanning. After the powder is laid, a high energy density laser is used to melt and fuse selected areas according to the processed data. Once the laser scanning is completed, the building platform is lowered, a next layer of powder is deposited on top and the laser scans a new layer till the completion of part [89, 3] as shown in figure 3.1. The building of parts starts from laying a thin layer of metal powder on a substrate plate in a building chamber. It is generally provided with inert gas to protect the heated metal parts against oxidation. This allows the manufacturer to form the necessary functional parts with the required benefits.

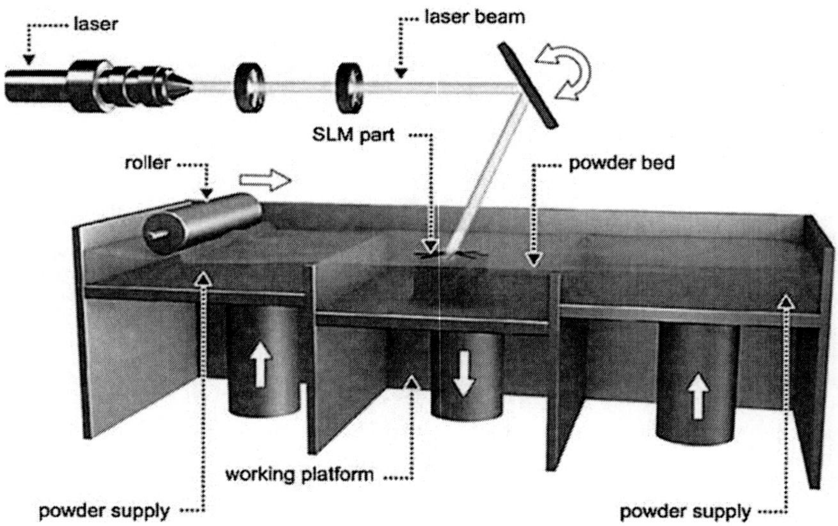

Figure 3.1: Schematic Process of SLM [3]

3.4. Features Attained Using SLM Process

As in any conventional process, the feature resolution, surface finish, mechanical properties, packing density, porosity and microstructure are characterized by SLM process as shown in the figure 3.2 in order to evaluate the quality of the final built part by the SLM process. These are the characteristic output for SLM process where various parameters are examined and the results are provided basis on the output characteristics of SLM process.

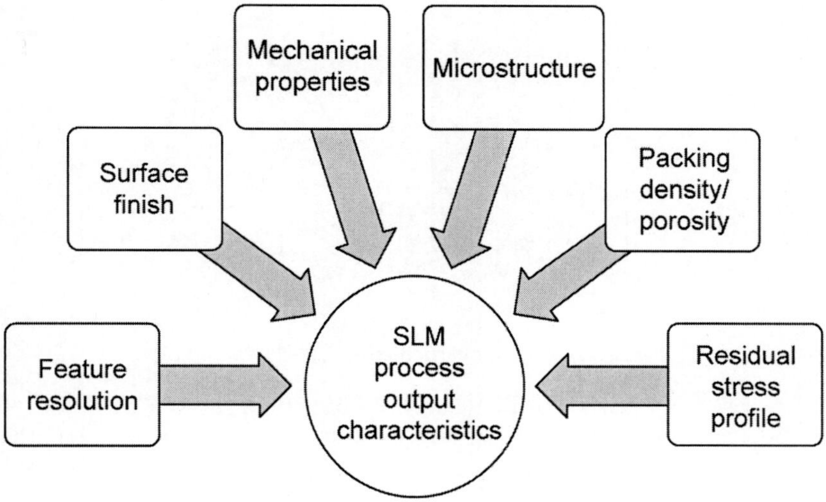

Figure 3.2: Output Characteristic of SLM [4]

3.5. Components of SLM Machine

EOS GmbH, Krailling, Germany is a private-owned machine manufacturer that sees itself as a solution provider with a strong background in AM. The current EOS metal printing machine M280/290/400 is a laser sintering/melting machine, with which it is possible to process different metal powders. In addition to various steels, titanium alloys, nickel based super alloys, and aluminium alloys are used for production of components. Theoretically, every weld-able material can be processed by these systems; however, the availability of exotic materials as defined metal powders in constant qualities is limited at the moment.

This type of machine as shown in figure 3.3 shows SLM 280 which is equipped with a 200/400 IPG -fiber laser. It combines the advantages of high beam quality, high beam intensity, and high dynamics. The machine is composed of a laser, optical unit, scanner, process computer, process chamber, and inert gas re-circulating filter system. The building process is operated under inert gas atmosphere such as nitrogen or argon or Helium.

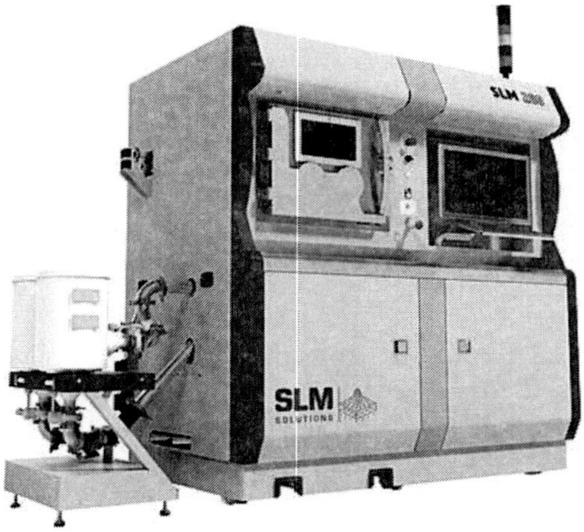

Figure 3.3: Schematic Illustration of SLM 280 Machine [5]

For controlling the building process, several sensors are available. The systems are controlled by standard industry computers where the control of machine components, surveillance, scanning patterns, and scanner/laser communications are made. The central point of the process chamber is the elevator system. These components are the building-, dispenser-, and collector platforms. The build platform lowers while the manufacturing process stepwise depending on the layer thickness and is performed after every exposed layer as shown in figure 3.4.

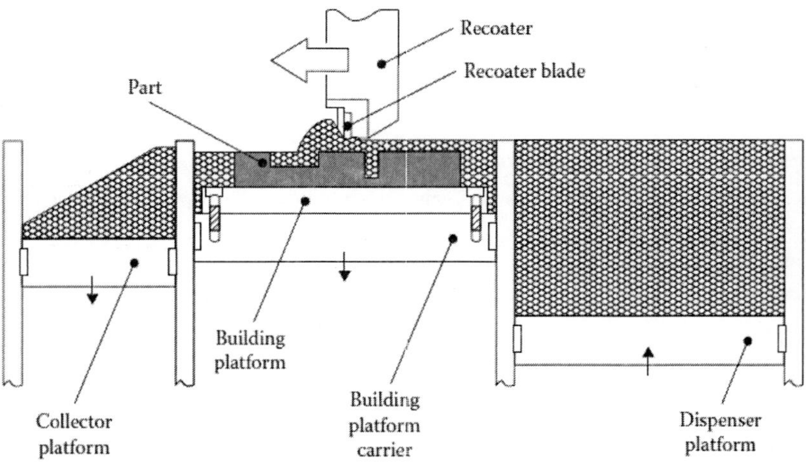

Figure 3.4: Functional Description of the Building Process [6]

On the right hand side of the building platform, the dispenser platform is located. (It is also possible to feed fresh powder from above.) It feeds fresh metal powder for the manufacturing process. The powder is deposited from the dispenser over the build platform via a recoater arm. The recoater applies the required amount of powder onto the building platform in the adjusted layer thickness. On the left hand side, the collector system is positioned. It collects the metal powder overflow and process side-products of the building process. Like the building platform, it is lowered after several recoating.

The building platform carrier is responsible for movements in z-height and thus the required layer thickness of the part. Onto the platform carrier, a replaceable metallic platform is fixed. This metal platform (generally made from the same material as the currently used powder or materials with similar thermal expansion coefficients) acts as a substrate and has two major functions. It first acts as a heat-sink to absorb the high energy concentrations from the melt pool and second as a bonding substrate for the metallic components printed on top. It holds the parts on position and compensates internal stresses from the process that could cause warpage or delamination off the platform.

After the process is finished, the platform together with the printed components is removed from the machine. In general, a stress-relieving heat treatment is performed, and the parts are then separated from the platform by wire EDM or band saw cutting. This platform can be easily recycled after component removal by just milling/grinding it flat again. It is also possible to build hybrid components, so for instance a preform can be mounted to the platform and only complex features are added on top. Part repair is also possible by this technology on turbine blades, gas turbine injection nozzles, among others.

3.6. Factors to be Considered in SLM

There are many factors which have to be considered for manufacturing components having unique, complex geometries and customization to enable industrialized implementation. The associated aspects such as process, design and material related factors are discussed. The process related and other related factors are depicted in figure 3.5 and in table 3.1respectively.

The design related factors concerns about the optimization of product design (Topology Optimization, TO) and sustainability in terms of reducing weight without the issues while fabricating the product. While the process related factors are considered for optimizing the process for fabricating the product with respect to material related factors.

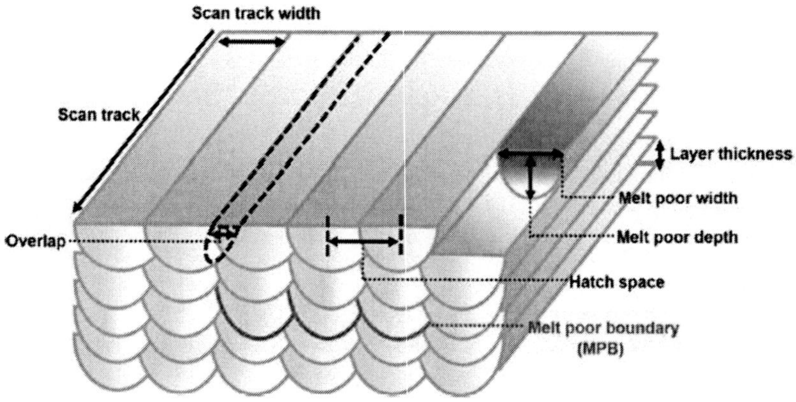

Fig 3.5: Process Related Factors of SLM Process [7]

Table 3.1: Significant Process Related Factors of SLM Process [8]

Powder related properties(Material)	**Laser related properties**	**Powder bed related properties**
Chemical composition	Laser type	Deposition strategy
Particle size and form	Laser intensity	Depth of particle layer
Spreading of Particle	Spot size	Powder bed density
	Scanning rate	Powder bed temperature
Powder flowability	Hatch spacing	
	Scanning strategy	Building chamber atmosphere
Apparent density		
Absorptivity		

3.6.1. Process Related Factors

The process related factors are categorized into Laser related, powder related and powder bed related properties. The preceding section will deal about the related factors such as power and type of laser used, Scanning speed, layer thickness and line or hatch spacing. It should be noted that most of these parameters are strongly interdependent and are mutually interrelated.

3.6.2. Laser Related Factors (Laser Power, Scanning Speed, Layer Thickness, Scan Line Spacing)

In general, laser processing of materials, specifically in metals is a controlled heating process where the laser acts as a source of controllable heat because of its energy. In the case of photopolymers, the absorbed laser light polymerizes or hardens the area of the polymer that it irradiates, producing a solid part. When a laser beam impinges on the surface of a material either a solid powder, or liquid some light is reflected, some is absorbed, and some may be transmitted, depending on the optical properties of the work piece. Laser output power, spatial distribution, time dependence (pulsed or continuous), beam diameter, and beam divergence affect how well the beam can be controlled in terms of the rate at which energy can be delivered to the material and how this energy is distributed spatially. Spatial distribution [9, 10] of the laser beam on the surface of a material affects the maximum irradiance obtained and the resultant temperature distribution.

The SLM process is set by adjusting various parameters. An optimal combination of laser power, scanning speed, powder layer thickness, and scan line spacing is required to minimize the potential defects by optimizing the process factors to produce high quality parts. In order to improve the precision, an offset i.e. about the radius of the laser beam, a fill contour can also be used for a higher precision or an improved outer boundary quality as shown in figure 3.6.

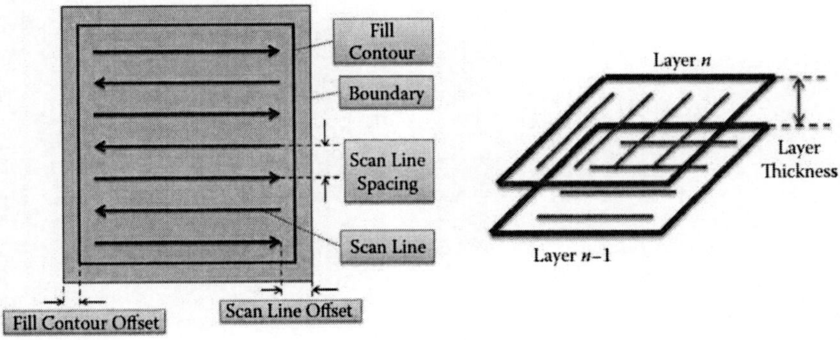

Figure 3.6: Scanning Pattern for SLM Process [11]

The effect of the process parameters can be combined and presented as the energy density, which is an engineering parameter representing the energy delivered to a unit volume of powder material. This is achieved by combining the laser power, scanning speed, scan line spacing, and layer thickness as shown below.

$$E_\rho = \frac{P}{vst} \ (J/mm^3) \qquad (3.1)$$

Where,

E_ρ = energy density (J/mm3 or J/mm^2),

P = laser power (W),

v = scanning speed (mm/s),

s = scan line spacing (mm), and

t = layer thickness.

From the equation 3.1 it is clear that increasing the laser power and decreasing the scanning speed, scan line spacing, or layer thickness increases the laser energy density.

Laser with a fiber, diode, with a wavelength around 1 μm is more efficient compared with processing with a CO_2 laser operating at 10 μm because of greater metal absorptivity at the shorter wavelength [12].

3.6.3. Scanning Strategies

Different scanning strategies as shown in figure 3.7 such as A - unidirectional scanning, B - bidirectional scanning, C - re-scanning based on bidirectional scanning with 90° rotation of the scanning direction, D - island scanning, E - shifting the position of the scan vectors between the layers based on island strategy in transverse direction(TD), Building direction(BD) and Scanning directions(SD) respectively which affects the SLM process and consequently alter the material properties, including density, thermal and residual stress, and microstructure.

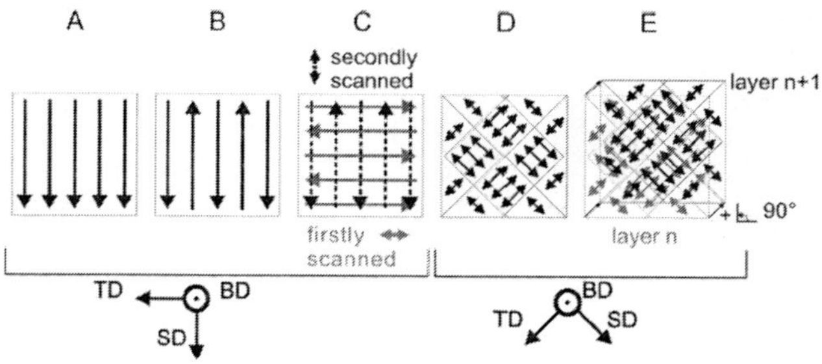

Figure 3.7: Scanning Strategies in SLM Process [13]

It has been reported, that an alternating scanning strategy (rotating the scanning direction in each layer) improves the density [13]. Moreover it has an influence on residual stresses and can be modified to mitigate quality issues such as warpage, cracks, and delamination etc.

3.6.4. Material Related Factors

The morphology and size of feedstock powders are important factors for powder bed fusion processes because they affect powder flowability, laser energy absorption and the thermal conductivity of the powder bed. A higher packing density of the powder bed is preferred for powder bed fusion processes because of the lower internal stresses, part distortion, porosity and surface roughness in the built part. The packing density of the powder bed is significantly affected by the particle morphology, size and distribution [14]. A higher powder bed packing density can be achieved with a wider distribution of particle size, as the gaps between larger particles can be filled in with smaller particles.

The spherical particle morphology improves the flowability of the powder to achieve a high packing density in the powder bed, which improves the final quality of the SLM-processed parts. Hence gas-atomized powders with a spherical shape are widely used as the feedstock for additive manufacturing processes. On the other hand, the flow of powder with a non-spherical shape is obstructed because the particles tend to interlock mechanically and entangle with each other. Hence non-homogeneous layers of powder with varying packing densities form on the top of the previously built solid surface, which may lead to the formation of defects such as porosity and/or incomplete melting.

A reduction in particle size results in (1) an increase in surface area, which favours the absorption of laser energy to increase the melt pool temperature; (2) an increase in the gap in the powder bed, which may lead to high porosity in the consolidated part if the gap is too large [15]; (3) an increase in the tendency for particle agglomeration [16]; and (4) a reduction in powder flowability.

The thermal conductivity of loose powder is primarily determined by the packing density of the powder bed and, less significantly, by the properties of the powder material. In general, the thermal conductivity of powder beds formed by irregular particles or by those with a wide size distribution is higher than that formed by monodisperse spherical powders at the same density [17].

The incident laser power is absorbed by the powder layer by multiple reflections through the gaps among the particles. Hence the absorption of laser energy by the powder bed is significantly higher than the absorption by a flat surface of a solid metal with the same composition. The effect of multiple reflections on absorption is more significant for highly reflective metals (such as aluminium and copper) than for moderately absorptive metals (such as iron and titanium); therefore the dependence of absorptivity on powder array and powder feed system is more pronounced for highly reflective metals. The percentage of the incident energy absorbed by the substrate or the previously built solid underneath the powder layer decreases with an increase in the optical thickness of the powder layer and a decrease in particle size.

3.6.5. Powder Bed Related Factors

In the early days of the technology [5], low-power of 50-100 W CO_2 and Nd:YAG lasers with focused spot sizes of 200-300 mm in diameter were used as heat sources to melt the powder and manufacture parts. The machine build volume was also relatively small. The growth of the technology in various fields has driven developments in beam scanning technology and platform/volume size, with the largest commercial volumes today used with a high power laser which in turn gives the powder bed temperature during the process.

In addition to increasing the volume size and laser power to increase the build rate, a new build strategy has been developed because of the increase in the output power of laser sources. This involves manufacturing a number of the surfaces of the part layers, first using a finer powder layer thickness, typically 30-50 μm, followed by a single core layer thickness depending on the total number of layers. This allows the part to be manufactured with a relatively low surface roughness but at a higher speed.

The atmosphere in building the parts in SLM with the use of Argon, Helium and Nitrogen gases which can also have an impact on the properties of parts being manufactured or produced[18, 19].

3.7. General Considerations for Supports Design in Metal SLM

When components are not specifically designed for SLM, supports are critical to ensure fabrication success as well as high and efficient production yield. Optimising build orientation is the 1st step to minimise the amount of supports by finding. Choosing the most appropriate type of support is the second step. These supports need to be strong yet easy to remove while using minimal material.

3.7.1. Part Geometry

As noted, SLM can produce complex geometries and internal structures with no impact to manufacturing cost, and so encourages the design of freeform models with optimized topology, shape, and/or size. However it is useful to consider the fundamental elements of the SLM process during this initial model generation and optimization. SLM is a powder-bed fusion process, and it is therefore prudent that parts be designed to eliminate powder entrapment and simplify powder removal. The size and geometry of the laser and melt pool should be considered during initial design, as these identify the minimum wall thickness and feature resolution achievable e.g. the circular cross-section of the incident laser causes the rounding of edges. SLM parts are manufactured layer-wise, and accordingly tend to be approximations of the original model since each layer must be of finite thickness. One consequence of this is that features (e.g. surfaces, bores, gaps) not aligned with the direction of printing will be impacted by the staircase effect. This can result in undesirable surface finishes, dimensional inaccuracies, and feature losses. The impact of the staircase effect can be minimized through careful design and designing the part to facilitate subsequent post-processing operations.

3.7.2. Build Orientation

The chosen build orientation is critical as it impacts the location of support structures, the manufacturing time, and ultimately the quality of the produced part (surface quality, dimensional accuracy, microstructure, and mechanical properties). While the optimal orientation is case dependent, it is generally wise to orient parts to minimize the number of surfaces that require support and to prevent the supporting of functional and hard-to-reach / internal surfaces. An orientation should also be selected that minimizes the deviation in the part's cross-sectional area (CSA) between succeeding layers, as sudden changes generate high temperature gradients and residual stresses that can cause part distortion and cracking. This is due to the low heat dissipation between the melted layer and underlying powder, which highlights the need to consider different heat transfer scenarios. When deciding upon part orientation. SLM is an anisotropic build process that produces parts with a notable difference in mechanical properties and microstructure along the axis parallel to the build direction. As-built parts have a poorer UTS and ductility in this direction, and so the intended use of the part should be considered during orientation selection. Overall build height is another factor which should be taken into account when orienting a part, as higher builds require significantly more time due to the increased number of layers. In addition, tall parts can lead to building failures, especially when other dimensions are comparatively smaller, as the part can become unstable during the layering process [20].

3.7.3. Supports

Support structures are required within AM as they provide the scaffolding and reinforcement required to ensure that a part and its features don't collapse due to gravity during the build process. In SLM, support structures are particularly important as they securely connect the part with the base plate. This direct connection impedes part displacement during powder deposition and provides a heat sink that helps to dissipate heat and reduce thermal stresses in the part. The goal during support structure generation is to try and achieve a structure which is strong enough to hold the part and prevent warping, but fragile enough to be easily removed after the build process is complete. Ease of removal is important, as it reduces post-processing times and can lead to better surface quality after support removal. One approach to obtaining secure but easily detachable supports is the use of teeth between the main support and part, as their spacing and geometry can be adjusted to achieve a suitable balance. A range of different support structure geometries are commonly used within SLM and can be automatically applied to a model within different software packages. However automatically generated supports tend to require manual redesign when dealing with complex shapes so as to improve their efficiency while minimizing their volume. When manually designing supports, the overhang angle of each downward facing surface should be considered (surface angle relative to the base plate). While geometry and material specific, a general guideline within SLM is that overhang angles of less than 45° must be supported in order to overcome the staircase effect and prevent deformation of the part.

A support structure is composed of two functional areas: the main support itself and the teeth. The teeth are the connections between the main support and the part. Their role is to minimise the contact area (and facilitate removal), act as efficient localised mini heat-sinks (for improved surface roughness) and fix the 1st overhangs layers in place (secure against warping).

The main peg must be strong enough to withstand upper vertical weights; horizontal recoating and thermal stresses. Teeth design can be fine tuned so as to define top length, based interval between teeth and teeth height: to close, and the teeth degrade the surface roughness; too far and they become too weak to be efficient.

There are various geometries available for the main support and teeth support as shown in figure 3.8 and 3.9. A point support is used for very small features, a web support for circular areas, a line support for narrow down facing areas. The most widely (and successfully) used is the block structure. Block supports are applied for large overhangs. Their design is based on a grid of lines fabricated along the x and y directions (in the substrate plane). The distance between lines, i.e. hatching, can be varied.

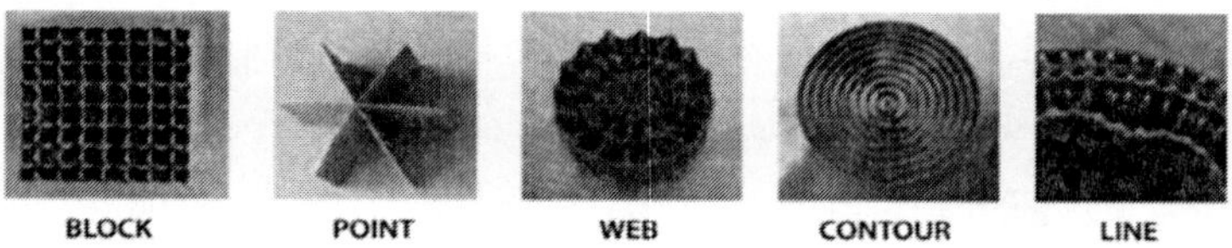

Figure 3.8: Various Types of Structure Geometries Used in SLM Process [21]

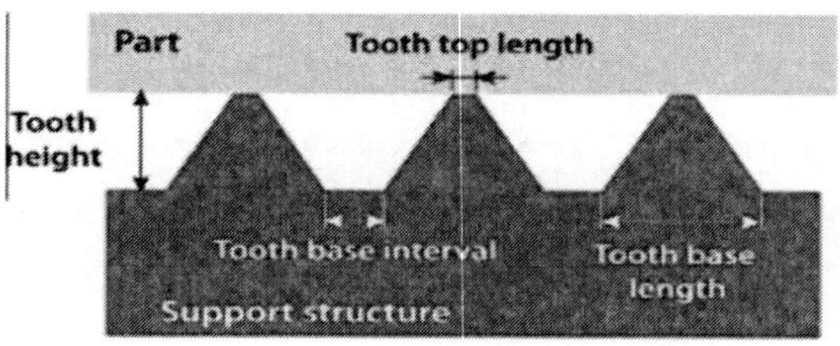

Figure 3.9: Typical Support Teeth Concept [21]

To facilitate supports removal, each block can in turn be fragmented and the pegs (main support) can be perforated with diamond or rectangular shapes.

3.7.4. Sacrificial Geometries

The software automatically generates the supports such as Magic's software. Their proprietary algorithms identify surfaces inclined over a user-defined angle threshold to build supports reaching towards the platform. However, it is sometimes more sensible to actually add sacrificial geometries: they will limit the amount of supports structure required or the time involved to remove them. A typical sacrificial geometry is, for instance, a platform oriented at 45° incidence with solid supports at each extremity. Something that would look similar to an extruded 'V' letter and that can generally be wire-cut quickly. This is especially viable when some very tall block supports are required. They are naturally prone to distortion (warping) and tend to increase the risk of unrecoverable fabrication failure.

Considerations such as build success probability, operator time and skill levels, all need to be taken into account when dealing with supporting structure. The quantity of supports impacts the long term commercial viability of the process. Teeth influence surface roughness. It is critical to build the component at an optimised angle to minimise the amount supports (hence material) or the post processing time. Using the right type of structure, most frequently fragmented blocks with hollows trunks and strong teeth, is also very important. Where components are not suitably (re)designed for selective laser melting, supports structure play a significant role in fabrication success and commercial viability.

References

[1] C.K. Chua and K.F. Leong, *"3D Printing and Additive Manufacturing: Principles and Applications"*, 2015, 4th edition, p. 518.

[2] Gu, D.D., Meiners, W., Wissenbach, K., & Poprawe, R, "Laser additive manufacturing of metallic components: materials, processes and mechanisms". *International Materials Reviews,* 2012, 57(3), 133–164.

[3] *"Implementing Additive and Cold Spray Techniques for the smart manufacturing of valve components (IMPACT)"*, Research, STAM - Science & Technology in Advanced Manufacturing.

[4] Nagarajan, B., Hu, Z., Song, X., Zhai, W., & Wei, J, *"Development of Micro Selective Laser Melting: The State of the Art and Future Perspectives"*, 2019, Engineering 5. Pg. 702–720.

[5] https://www.slm-solutions.com/products/machines/

[6] Bandyopadhyay, Amit Bose, Susmita, *"Additive manufacturing"* book, CRC Press, 2016, Taylor & Francis Group.

[7] Wang, J., Liu, S., Fang, Y., & He, Z, "A short review on selective laser melting of H13 steel", *The International Journal of Advanced Manufacturing Technology,* 2020.

[8] Gunasekaran, J., Sevvel, P., & John Solomon, I, "Metallic materials fabrication by selective laser melting: *A review. Materials Today: Proceedings,* 2020.

[9] R. Poprawe, K. Boucke, Tailored Light 1: *High Power Lasers for Production,* Springer, Berlin, 2012.

[10] G.G. Gladush, I. Smurov, *Physics of Laser Materials Processing,* Springer, Berlin, 2011.

[11] Dadbakhsh, S., Hao, L., & Sewell, N, "Effect of selective laser melting layout on the quality of stainless steel parts", *Rapid Prototyping Journal,* 2012, 18(3), pg. 241–249.

[12] Lee, H., Lim, C.H.J., Low, M.J., Tham, N., Murukeshan, V.M., & Kim, Y.J, "Lasers in additive manufacturing: A review. *International Journal of Precision Engineering and Manufacturing-Green Technology",* 2017, 4(3), pg. 307–322.

[13] Yu, W.H., Sing, S.L., Chua, C.K., Kuo, C.N., & Tian, X.L, "Particle-Reinforced Metal Matrix Nanocomposites Fabricated by Selective Laser Melting: A State of the Art Review". *Progress in Materials Science,* 104, 2019, pg. 330–379.

[14] Zhang, J., Gu, D., Yang, Y., Zhang, H., Chen, H., Dai, D., & Lin, K, "Influence of Particle Size on Laser Absorption and Scanning Track Formation Mechanisms of Pure Tungsten Powder during Selective Laser Melting", *Engineering. Additive Manufacturing Article, Engineering* 5, 2019, pg. 736–745.

[15] S. Kumar, *Selective laser sintering/melting, in: Comprehensive Materials Processing,* 2014, pg. 93e134.

[16] D. Manfredi, F. Calignano, M. Krishnan, R. Canali, E.P. Ambrosio, E. Atzeni, "From powders to dense metal parts: characterization of a commercial AlSiMg alloy processed through direct metal laser sintering", *Materials* 6, 2013, pg. 856-869.

[17] Olakanmi, E. O, "Selective laser sintering/melting (SLS/SLM) of pure Al, Al–Mg, and Al–Si powders: Effect of processing conditions and powder properties", *Journal of Materials Processing Technology,* 2013, 213(8), pg. 1387–1405.

[18] Wang, X.J., Zhang, L.C., Fang, M.H., & Sercombe, T.B, "The effect of atmosphere on the structure and properties of a selective laser melted Al–12Si alloy", *Materials Science and Engineering: A* 597, 2014, pg. 370–375.

[19] Salehi, M., Maleksaeedi, S., Farnoush, H., Nai, M.L.S., Meenashisundaram, G.K., & Gupta, M, "An investigation into interaction between magnesium powder and Ar gas: Implications for selective laser melting of magnesium", *Powder Technology,* 333, 2018, pg. 252–261.

[20] Zhang, S., Rauniyar, S., Shrestha, S., Ward, A., & Chou, K, "An experimental study of tensile property variability in selective laser melting". *Journal of Manufacturing Processes* 43, 2019, pg. 26-35.

[21] https://www.insidemetaladditivemanufacturing.com/blog/considerations-for-supports-design-in-metal-slm

Chapter IV

Selective Laser Melting (SLM)

of

Aluminium (al) Alloys

CHAPTER IV

4. Selective Laser Melting (SLM) of Aluminium (Al) Alloys

4.1. Introduction

Aluminium is the world's most abundant metal and is the third most common element, comprising 8% of the earth's crust. The versatility of aluminium makes it the most widely used metal after steel. Pure aluminium is soft, ductile, and corrosion resistant and has a high electrical conductivity. It is widely used for foil and conductor cables, but alloying with other elements is necessary to provide the higher strengths needed for other applications. Commercial heat treatable Al alloys are 2000 (Al–Cu or Al–Cu–Mg), 6000 (Al–Mg–Si), and 7000 (Al–Zn–Mg) series alloys whose strength and other properties are enhanced by various heat treatment processes whereas non-heat-treatable Al alloys include 1000, 3000, 4000 containing only Si, and 5000 series alloys which are strengthened by mechanical deformation processes [1-3].

Engineering parts made in aluminium alloys have been fabricated by traditional manufacturing processes such as casting, forging, extrusion, and powder metallurgy (P/M). However, these traditional manufacturing processes have resulted in parts having coarse grain structures with the attendant poor mechanical properties as a consequence of low cooling rates associated with these processes. Moreover, the adoption of tooling for making Al alloy parts through these traditional manufacturing routes increases the cost of production and the lead time. Therefore, for manufacturing businesses engaging in the fabrication of aluminium parts and aiming to deliver their new customised products more quickly and gain more consumer markets for their products.

The major advantages of using aluminium are based on their remarkable properties [4]. Some of these properties are outlined in the subsequent sections.

4.1.1. Strength to Weight Ratio

Aluminium has a density around one third that of steel and is used advantageously in applications where high strength and low weight are required. This includes vehicles where low mass results in greater load capacity and reduced fuel consumption.

4.1.2. Corrosion Resistance

When the surface of aluminium metal is exposed to air, protective oxide coating forms almost instantaneously. This oxide layer is corrosion resistant and can be further enhanced with surface treatments such as anodising.

4.1.3. *Electrical and Thermal Conductivity*

Aluminium is an excellent conductor of both heat and electricity. The great advantage of aluminium is that by weight, the conductivity of aluminium is around twice that of copper. This means that aluminium is now the most commonly used material in large power transmission lines. The best alternatives to copper are aluminium alloys in the 1000 or 6000 series. These can be used for all electrical conduction applications including domestic wiring.

Weight considerations mean that a large proportion of overhead, high voltage power lines now use aluminium rather than copper. They do however, have a low strength and need to be reinforced with a galvanised or aluminium coated high tensile steel wire in each strand.

4.1.4. *Toxicity*

Aluminium is not only non-toxic but also does not release any odours or taint products with which it is in contact. This makes aluminium suitable for use in packaging for sensitive products such as food or pharmaceuticals where aluminium foil is used.

4.1.5. *Recycling*

The recyclability of aluminium is unparalleled. When recycled there is no degradation in properties when recycled aluminium is compared to virgin aluminium. Furthermore, recycling of aluminium only requires around 5 percent of the input energy required to produce virgin aluminium metal.

The combination of two remarkable properties of aluminium makes the need to recycle the metal obvious. These first of these factors is that there is no difference between virgin and recycled aluminium. The second factor is that recycled aluminium only uses 5% of the energy required to produce virgin material. Currently around 60% of aluminium metal is recycled at the end of its lifecycle but this percentage can still be vastly improved.

4.1.6. *Applications*

Aluminium is one of the lightest engineering metals, having strength to weight ratio superior to steel. By utilising various combinations of its advantageous properties such as strength, lightness, corrosion resistance, recyclability and formability, aluminium is being employed in an ever-increasing number of applications. This array of products ranges from structural materials through to thin packaging foils.

4.2. Densification Behaviour based on Laser Related Process Parameter

Ghosh et al. [5] investigated an in-situ multi-component reinforced aluminium based metal matrix composite, consisting of different mixtures of Al, TiO_2 and B_4C powders, fabricated by the combination of self-propagating high-temperature synthesis and SLS processes as layer thickness, laser power, scanning speed, hatch spacing and powder composition were varied in an attempt to optimise for getting higher density, lower porosity using Taguchi design of experiments (DOE). Three homogeneous powder mixtures of Al (99.5% purity, −325 mesh), TiO_2 (anatase, 99.9% purity, −325 mesh) and B_4C (100% purity, maximum particle size — 4 μm) powder were prepared. The amounts of Al in the mixtures were 80 wt.%, 85 wt.%, and 90 wt.% and the remaining 20 wt.%, 15 wt.%, and 10 wt.% respectively composed of TiO_2 and B4C in a molar ratio of 3:1.

Table 4.1: Orthogonal Array L18 [5]

Specimen no.	**Layer thickness (μm)**	**Powder composition (wt.% of Al)**	**Laser power (W)**	**Scan speed (mm/min)**	**Hatching spacing (μm)**
1.	200	80	600	2000	0.4
2.	200	80	800	4000	0.5
3.	200	80	1000	6000	0.6
4.	200	85	600	2000	0.5
5.	200	85	800	4000	0.6
6.	200	85	1000	6000	0.4
7.	200	90	600	4000	0.4
8.	200	90	800	6000	0.5
9.	200	90	1000	2000	0.6
10.	300	80	600	6000	0.6
11.	300	80	800	4000	0.4
12.	300	80	1000	2000	0.5
13.	300	85	600	4000	0.6
14.	300	85	800	6000	0.4
15.	300	85	1000	2000	0.5
16.	300	90	600	6000	0.5
17.	300	90	800	2000	0.6
18.	300	90	1000	4000	0.4

The SLM process is repeated till the required sintered specimen height is obtained with laser spot size of around 700 μm could be obtained on the powder bed. By varying a scanning rate of 2000,4000 and 6000 mm/min, laser power of 600,800,and 1000 W and hatch spacing of 0.4, 0.5 and 0.6 μm as described in table 4.1 stating layer thickness as 200 μm and 300 μm. The experiments were conducted as per L18 orthogonal array specified by varying the specified laser related process parameters.

The results of the ANOVA for density are reported in Table 4.2. By comparing the F_{test} values with the F_{tab} values, it is found that the layer thickness, powder composition and hatching distance are significant factors. The contributions of these parameters on density are 20.14%, 41.83% and 36.66% respectively. Hatching distance also has a good influence on the density.

Table 4.2: ANOVA for Density [5]

Factors	**degree of freedom (DOF),**	**Sum of squares (SS),**	**mean squares (MS)**	**percentage of contribution (C)**	**F-test values (F_{test}).**	**F-test values (F_{tab}).**
Layer thickness (μm)	1	2.179401	2.179401	4.74	39.81	5.99
Composition (wt.% Al)	2	27.66539	13.83269	60.16	252.67	5.14
Laser power (W)252	2	15.33633	7.668163	33.35	140.07	5.14
Scan speed (mm/min)	2	0.286528	0.143264	0.62	2.61	-
Hatching distance (μm)	2	0.078583	0.078583	0.17	0.71	-
Error	8	0.43796	0.43796	0.95	-	-
Total	17	45.98419	45.98419	-	-	-

In order to know the quality characteristics and Signal-to-Noise (S/N) Ratio, Taguchi method is used [6, 7], the response is considered as one of the three quality characteristics.

The quality characteristics, which are the sense of desirability of the result, can be classified into three categories namely (i) nominal-the-best, (ii) smaller the- better, and (iii) larger-the-better. In "nominal-the-best" quality characteristic, the value of the result is non-zero and finite. In "larger-the-better" quality characteristic, the larger magnitude of the result will be preferred to the smaller ones. In "smaller-the-better" quality characteristic, the smaller value of the result is always preferred to the others.

In this investigation, density and micro hardness were considered as "larger-the-better" quality characteristic and porosity was considered as "smaller-the-better" quality characteristic. Therefore, the signal-to-noise ratios (S/N ratios) for density and micro hardness were calculated using the following equation:

$$S/N \text{ ratio} = -10 \log_{10} (\text{mean square reciprocal of responses}) \qquad (4.1)$$

The S/N ratios for porosity were calculated using the following equation:

$$S/N \text{ ratio} = -10 \log_{10} (\text{mean square of porosity}) \qquad (4.2)$$

Thus by analysis of variance by calculating the determined based on the degree of freedom(DF), sum of squares (SS), mean squares (MS), percentage of contribution (C) and F-test values (Ftest) [8, 9, 10]. The parameter which has greater F-test value than the F-tabulated value (Ftab) is the significant factor and ten compared the F_{test} values with the F_{tab} values, it is found that the layer thickness, powder composition and hatching distance are significant factors. The contributions of these parameters on density are 20.14%, 41.83% and 36.66% respectively. So, the powder composition has more influence on density as compared to the other two significant parameters. This can be attributed to the fact that when the weight (wt.) percentage of Al is decreased, wt.% of TiO_2 and B_4C mixture is increased. Accordingly, more amount of TiO_2 and B_4C participates in the reaction which creates a greater amount of reinforcements, such as Al_2O_3, TiC, and TiB_2.

As the density of these reinforcements are higher than the theoretical density of the composite, the overall density of the sintered composite increases. Hatching distance also has a good influence on the density. The increase in hatching distance leads to more pores formations, which decrease the density of the specimen.

For a particular laser energy input, if the layer thickness of powder material is increased, then the laser energy will not be sufficient to melt the bottom portion of the layer completely which may also encourage pore formation [11].

Main effects plot for density is shown in figure 4.1. It is found from the plot that a layer thickness of 200 μm, composition containing 80 wt.% of Al, laser power of 800 W, scan speed of 2000 mm/min and a hatching distance of 400 μm are the optimum parameters to obtain a higher density within the range of study.

The confirmation tests were carried out based on the optimal parameters for density. The predicted values of density were determined which show that the predicted density value 29.73 g/cm³ is near to the experimental value which is 2.75 g/cm³.

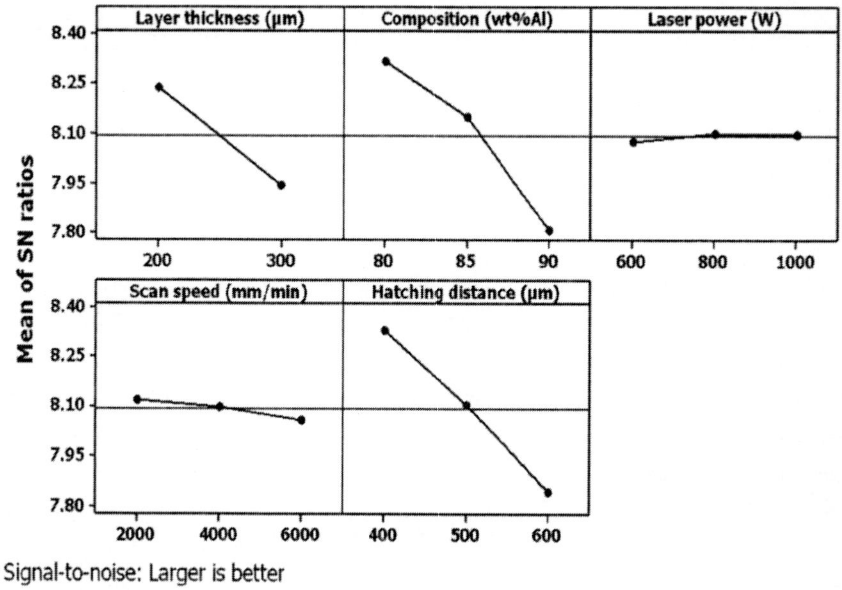

Figure 4.1: S/N Ratio for 800W, 2000mm/min and 400μm for Density [5]

Buchbinder and his team [12] investigates the main influencing variables of the primary process time are layer thickness (Ds), scanning velocity (vs) and hatch spacing (Δys). The process related build rate is calculated according to the following equation:

$$V_{Process} = D_S . V_S . \Delta y_S \quad (4.3)$$

Scanning velocity and layer thickness are limited by the laser power available whilst hatch spacing is limited by the focus diameter (Δys, max typically equals approx. 0.7 times the beam diameter).

SLM machines provide laser power only up to 200 W (400 W) a new SLM machine is newly designed, both in terms of hardware and software, to contribute to an increased process efficiency. In order to increase the process related build rate significantly beyond 5mm^3/s and the maximum laser power is extended to 1 kW. The laser source used within this setup is a fibre coupled disk laser. The beam profile is Gaussian shaped.

The flanks of the beam profile are steeper than those of a pure fundamental mode, whereas the top intensity is less, complanate than a top hat beam profile. The focus radius measures 97.7 µm and M^2 is calculated to 6.17 as shown in figure 4.2.

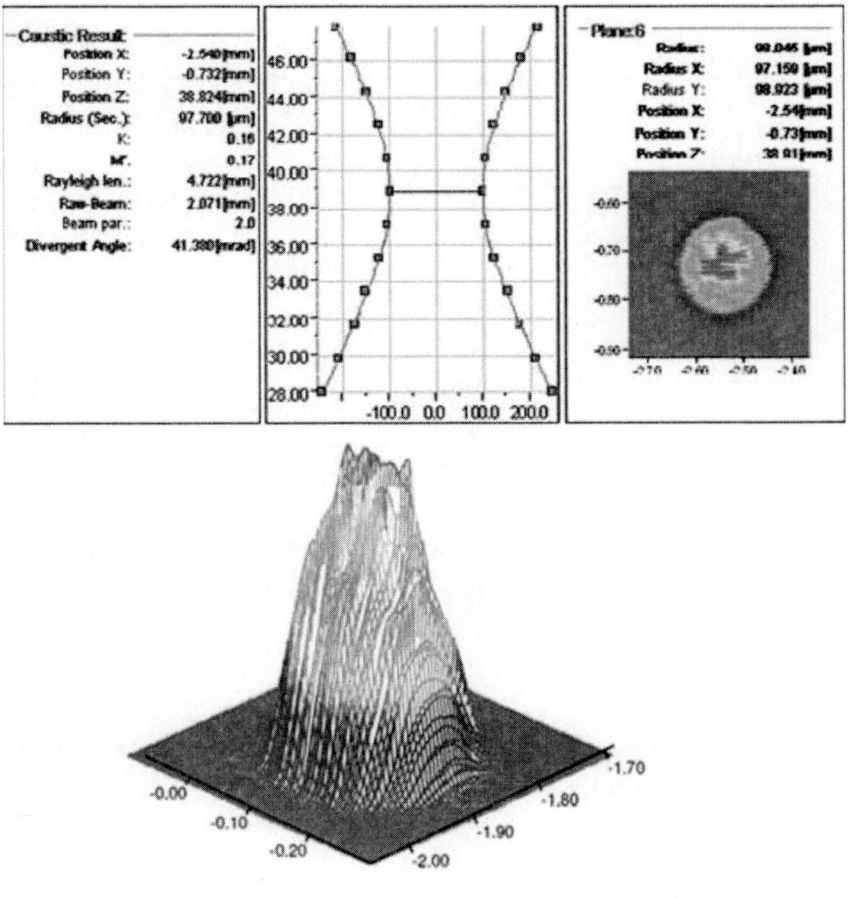

Figure 4.2: Caustic and Beam Intensity Distribution Measurement [50]

In order to build dense components, it is important that the melt pool does not consist of different subareas but is cohesive and self-contained. On the one hand the energy per unit length decreases with an increase of the scanning velocity. On the other hand the productivity of the SLM process is positively influenced by the scanning velocity. Hence, the laser power has to be increased if the scanning velocity increases. However, the augmentation of the laser power at a constant beam diameter increases the intensity at the point of processing which causes superheating and thus spattering and process instabilities. The high reflectivity and thermal conductivity of aluminium require a laser power of at least 300 W at a scanning velocity of 500 mm/s (build rate amounts to approx. 4 mm^3/s) to achieve densities approaching 100% for a layer thickness of 50 μm as shown in figure 4.3.

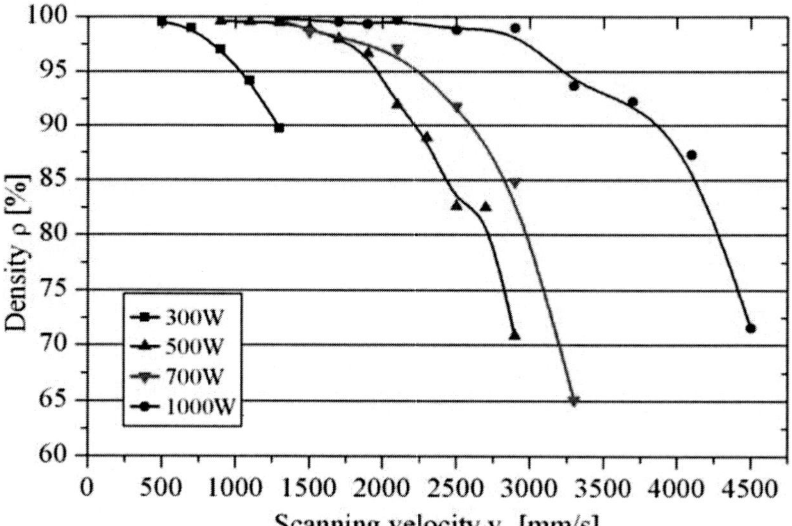

Figure 4.3: Density Depending on Scanning Velocity and Laser Power and Δy_s = 0.15 mm [12]

Considering a laser power of 500 W the production of components with a density of approx. 100% is possible at a scanning velocity up to 1200 mm/s. With regard to the current state of the art for 50 μm layers, i.e. 500 mm/s an increase of around 200% (9mm^3/s) can be achieved. If the scanning velocity is further increased the energy per unit length is not sufficient anymore to fabricate dense components. Regarding a laser power of 700 W dense components can be manufactured at a scanning velocity up to 1600 mm/s. Hence, the build rate can be augmented by more than 300% (12mm^3/s) considering the current state of the art. At a laser power of 1000

W, the scanning velocity can be increased to approx. 2200 mm/s which consequently increase the build rate (approx.16 mm^3/s). Figure 4.4 shows the density at different scanning velocities and laser powers is exemplified by means of cross sections.

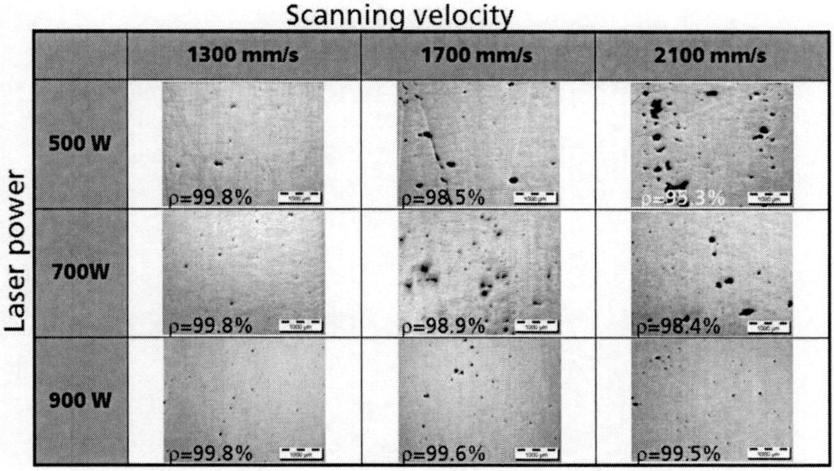

Figure 4.4: Densities by means of Cross Sections of SLM Samples Dep. on Scanning Velocity and Laser Power [12]

Besides scanning velocity, hatch spacing plays a major role in achieving higher build rates (see equation 4.3). In order to improve the build rate the increase of hatch spacing has been enlarged.

The results shown in figure 4.5 shows that densities of approx. 99.5% are still reached with a hatch spacing of 0.2 mm for 900 and 1000 W laser power. For a hatch spacing of 0.25 mm the density of samples produced with 900 W remains at approx. 99.5% as well. The lower density (approx. 96.8%) of samples produced with 1000 W and a hatch spacing of 0.25 mm is caused by spattering and process instabilities, which create imperfections. The amount of imperfections increases, because the area of remelting decreases, with increased hatch spacing. For a hatch spacing of 0.3 mm samples with a density higher 99.5% could not be built.

This increase of hatch spacing from 0.15 to 0.25 mm raises the build rate by approx. 60%, while a density of at least 99.5% is achieved as shown in figure 4.6 (PL = 900 W, vs. = 1700 mm/s, Ds = 50 μm, build rate of approx. 21 mm^3/s). This investigation demonstrates that the material AlSi10Mg, showing a high thermal conductivity, can be processed with a hatch spacing higher than the beam diameter.

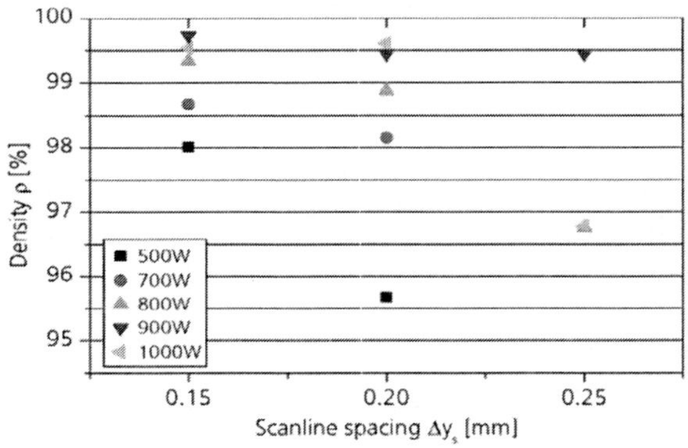

Figure 4.5: Density based on Process Parameters [12]

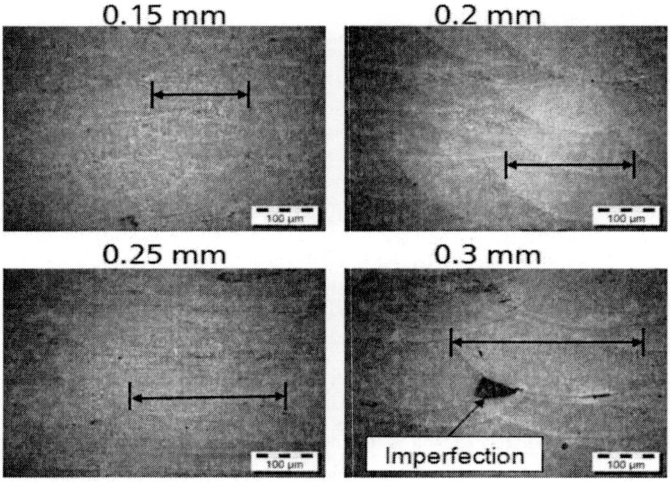

Figure 4.6: Cross Sections of Samples Built with different Hatch Spacing [12]

4.3. Densification Behaviour based on Powder Related Properties

Ghosh et al. [5] investigated the powder composition has more influence on density as compared to the other two significant parameters (i.e. layer thickness and hatch spacing) by varying the laser related factors. This can be attributed to the fact that when the weight (wt.) percentage of Al is decreased, wt. % of TiO_2 and B_4C mixture is increased as discussed in section 4.2. Accordingly, more amounts of TiO_2 and B_4C participate in the reaction which creates a greater amount of reinforcements, such as Al_2O_3, TiC, and TiB_2. As the density of these reinforcements are higher than the theoretical density of the composite, the overall density of the sintered composite increases.

4.4. Defects in SLM of Aluminium Alloys

4.4.1. *Porosity*

Porosity reduction by optimizing laser related process parameters was investigated by Ghosh et al. [5] by varying the laser related factors as discussed in section 4.2. The increase in hatching distance leads to more pores formation which decreases the density of the specimen. For a particular laser energy input, if the layer thickness of powder material is increased, then the laser energy will not be sufficient to melt the bottom portion of the layer completely which may also encourage pore formation [11]. The predicted porosity was calculated and the predicted values and experimental results obtained from the confirmation tests are merely equal 11.73 % and 11.5%.

The specimens were first placed in acetone bath and cleaned using an ultrasonic cleaning machine. The densities of the sintered specimen were measured repeatedly by the application of Archimedes' principle using precision balance (Make —Mettler Toledo, Model—AB265-S/FACT) of 0.03 mg least count.

The photographs of the top surface of the mirror polished specimen achieved using diamond paste (grade 1 and 1/4) were taken by an optical stereo zoom microscope (Make — Olympus, Model—SZ 1145 TRPT).The photographs as shown in figure 4.7 were then used for finding the porosities by digital image tool software. To get the average value of porosity of a specimen, the above process was repeated for three times.

It is evident from the ANOVA for porosity from table 4.3 that the hatching distance, layer thickness and laser power are significant parameters for porosity. Among these parameters, hatching distance which has a contribution of 72.81% is the major influencing parameter on porosity as compared to layer thickness (13.55%) and laser power (5.98%).

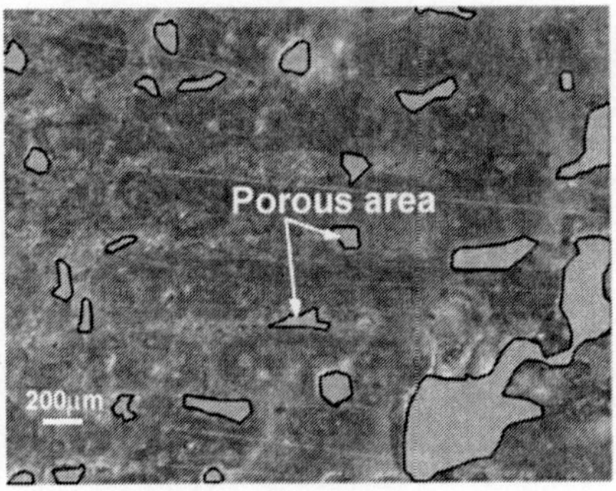

Figure 4.7: Schematic Image of the Porous Section [5]

Table 4.3: ANOVA for Porosity [5]

Factors	Degree of freedom (DF),	Sum of squares (SS),	Mean squares (MS)	Percentage of contribution (C)	F-test values (F_{test}).	F-test values (F_{tab}).
Layer thickness (μm)	1	3.211093	3.211093	13.55	28.38	5.99
Composition (wt.% Al)	2	0.048149	0.024074	0.20	0.21	-
Laser power (W)252	2	1.417706	0.708853	5.98	6.26	5.14
Scan speed (mm/min)	2	0.863383	0.431691	3.64	3.81	-
Hatching distance (μm)	2	17.2561	8.628049	72.81	76.26	5.14
Error	8	0.905071	0.113134	3.82	-	-
Total	17	23.7015		-	-	-

Increase in hatching distance and layer thickness increases the porosity. Since, S/N ratios for porosity were calculated based on "smaller-the-better" quality characteristic, so the higher average value of S/N ratio at a particular level of the factor in the main effects plot will be considered as optimum level for lower porosity. The main effects plot for porosity is shown in figure 4.8.

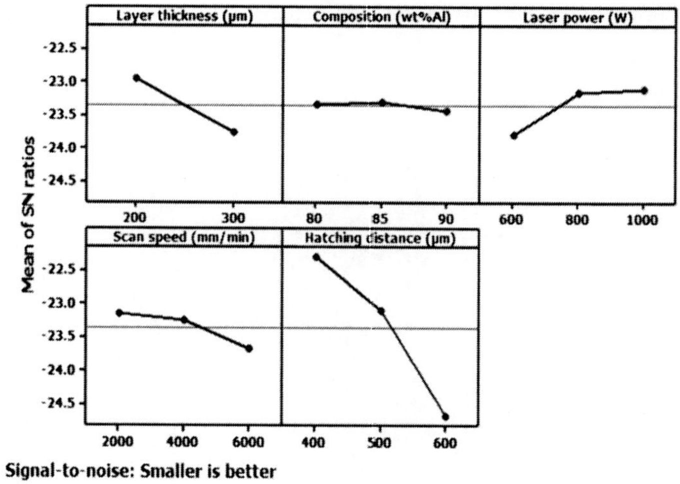

Figure 4.8: S/N Ratio with Respect to Process Parameters for Porosity [5]

4.5. Mechanical Properties of SLMed Part or Component

Microhardness prediction by optimizing laser related process parameters was investigated by Ghosh et al. [5] by varying the laser related factors as discussed in section 4.2. Generally a lower layer thickness, results into better melting of the powder material. This may lead to better compaction and higher microhardness. The parameters, other than composition, laser power and layer thickness has negligible effects on microhardness. The ANOVA for microhardness is furnished in Table 4.4. It has been observed that the composition, laser power and layer thickness are significant parameters.

Table 4.4: ANOVA for Microhardness [5]

Factors	Degree of freedom (DF),	Sum of squares (SS),	Mean squares (MS)	Percentage of contribution (C)	F-test values (F_{test}).	F-test values (F_{tab}).
Layer thickness (μm)	1	2.179401	2.179401	4.74	39.81	5.99
Composition (wt.% Al)	2	27.66539	13.83269	60.16	252.67	5.14
Laser power (W)252	2	15.33633	7.668163	33.35	140.07	5.14
Scan speed (mm/min)	2	0.286528	0.143264	0.62	2.61	-
Hatching distance (μm)	2	0.078583	0.039291	0.17	0.71	-
Error	8	0.43796	0.054745	0.95	-	-
Total	17	45.98419	-	-	-	-

The composition is the major influencing parameter and it has the highest contribution (60.16%) on microhardness as compared to laser power (33.35%) and layer thickness (4.74%). Though layer thickness is a significant parameter, it has the least contribution.

Since, S/N ratios for microhardness were calculated based on "larger-the-better" quality characteristic. So the higher average value of S/N ratio at a particular level of the factor in the main effects plot will be considered as optimum level for microhardness. The main effects plot for microhardness is shown in figure 4.9. It is found that a layer thickness of 200 μm, 80 wt. % of Al, laser power of 1000 W, scan speed of 6000 mm/min and hatching distance of 600 μm are the optimum levels to obtain better microhardness.

The confirmation tests were carried out based on the optimal parameters for microhardness. The predicted values of microhardness were determined which show that the predicted microhardness value of 1.74GPa and 1.7Gpa experimental value.

Buchbinder et al. [12] investigates the main influencing variables of the primary process time are layer thickness (Ds), scanning velocity (Vs) and scanline spacing (Δys) and also the effect on the properties such as hardness and tensile strength .Hardness is a first indication for the mechanical properties.

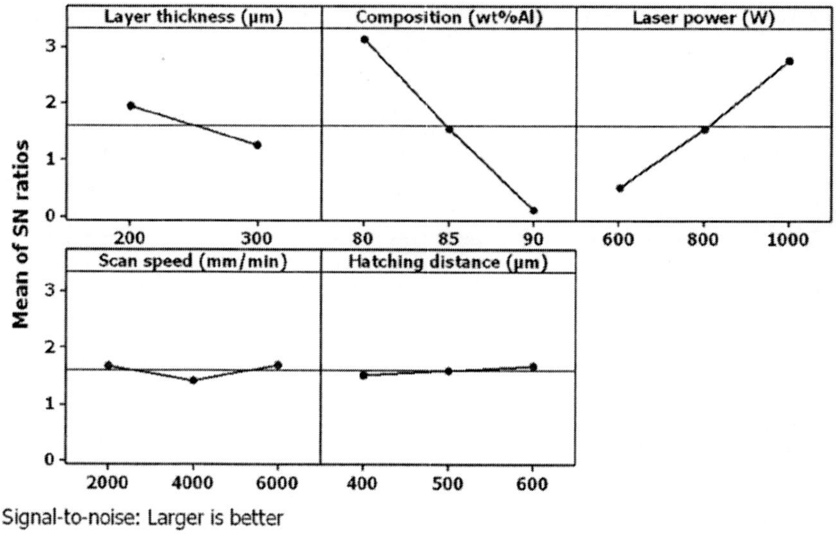

Figure 4.9: Effect of Microhardness on Laser Related Parameters [5]

Therefore, the hardness has been investigated for different process parameters. Figure 4.10 and 4.11 shows the resulting hardness of SLM samples at different scanning velocities and different laser power. The hardness of samples built with scanning velocities between 1000 and 4000 mm/s is approx. 145 HV 0.1. (The hardness is not depending on the density. The density for samples scanned with 4000 mm/s is approx. 90%).

The hardness is not influenced by using 500 or 900 W. In comparison to previous investigations [14], where an increase of hardness with higher scanning velocity has been demonstrated, the resulting hardness for this range of scanning velocity is not increased with higher scanning velocity. It seems that the hardness of the microstructure reaches a sort of maximum at approx. 145 HV0.1 and this already for scanning velocities of approx. 500 mm/s.

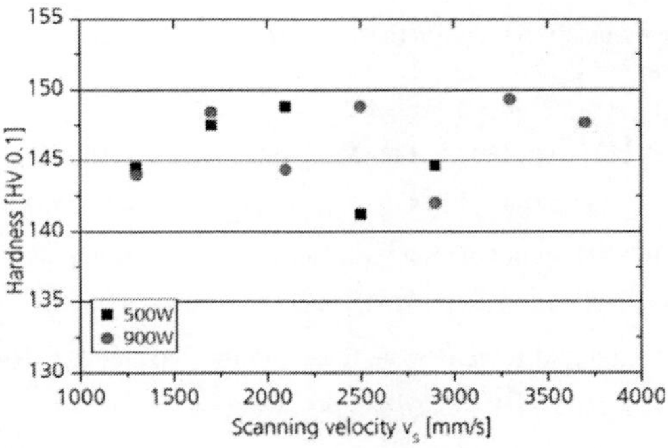

Figure 4.10: Hardness depending on Scanning Velocity [12]

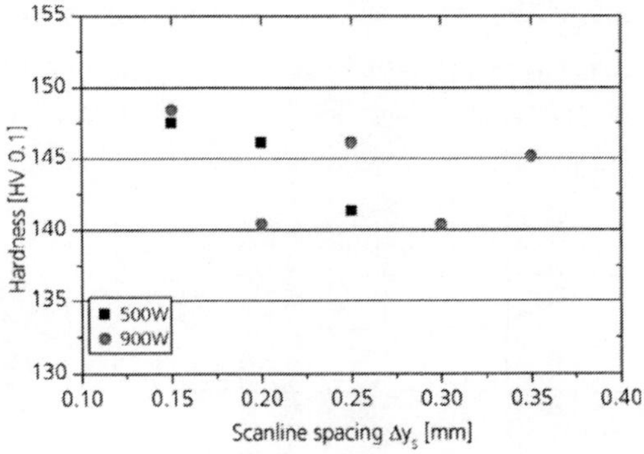

Figure 4.11: Hardness Depending on Hatch Spacing [12]

The hardness is not depending on the hatch spacing and reaches approx. 145 HV 0.1. As a result of the high scanning velocities and consequently high solidification rates the hardness of the SLM-made specimens is significantly increased by approx. 200% in comparison to the e.g. die casted parts. (According to EN 1706, Minimum hardness for AlSi10Mg is 74 HV). High hardness values indicate high tensile strength as well. In order to evaluate the mechanical properties tensile specimens were built according to DIN 50125 in three build-up directions (0°, 45°, 90° with reference to the applied load direction).

Figure 4.12 shows the tensile strength obtained with samples with a layer orientation parallel to the tensile direction (0°) show the highest tensile strength (R_m = 420 MPa), those built with a layer orientation perpendicular to the applied load (90°) show the lowest strength (R_m = 360 MPa). Independent from the build-up direction the measured values for the tensile strength outreach the minimum value (R_m = 240 MPa) according to EN 1706 for the die-cast material AlSi10Mg. The results of static tests so far demonstrate that the mechanical properties of the

SLM AlSi10Mg specimens obtain no less than the mechanical properties of serial-produced die-cast AlSi10Mg.

R.N. Lumley et al. [13] investigated the effect of particle size and evaluated the mechanical properties of Aluminium copper alloys. In this a binary Al-4wt%Cu alloy was used and the characteristics of the starting powders are given in table 4.5. All powders had a purity .99.5%. The powders were mixed with 0.5wt% stearic acid in a Turbula powder mixer for 20 minutes. Samples weighed 4–4.5g and were pressed in a floating rectangular shaped die (40 mm × ~ 9mm) at a pressure of 250MPa. Up to 16 samples were tested at each condition. Sintering was conducted in a high purity nitrogen atmosphere with a dew point of ≤ -40°C. Samples were either heated at 10°C/min or placed directly in a hot furnace to maximise the heating rate. The slow heating rate incorporated a 20 minute de-waxing step at 300°C. All samples were sintered at 600°C for 40 minutes and water quenched. The sintered bars were solution treated at 540°C for 1 hour, water quenched and aged at room temperature for 10–11 days (the T4 condition).

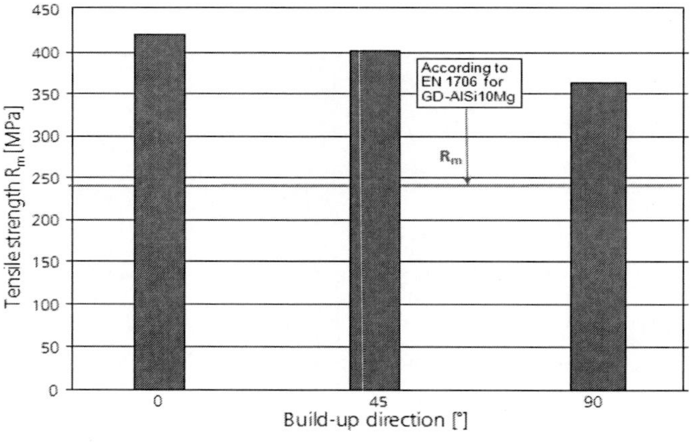

Figure 4.12: Tensile Strength based on Build up Direction [12]

Tensile samples were machined from the solution treated bars. The gauge dimensions were 20mm × 4.5mm × 2.5 mm. Tensile properties were determined using an Instron 1026 screw drive tensile machine with a 10kN load cell and a crosshead speed of 0.6mm/min. Strain was measured by an Instron model 2620/603 dynamic extensometer with a full scale range of 20%. The sintered density was measured following MPIF Standard 42, although ethanol was used instead of water.

Samples were prepared for micro structural analysis using standard polishing methods and finished on 0.05mm silica. These were etched in 0.5 vol% HF and examined optically.

Table 4.5: Powder Characteristics [13]

Powder	Description	Powder Sizes		Fabrication Method	Powder Shape
Aluminium	-	0-1%	-250µm+ 150µm	Atomised	Rounded
		35-55%	-150+63µm		
		Bal. 32-52%	-63+45µm		
			-45µm		
Copper	coarse	125-150 µm		Atomised	Spherical
	Intermediate	75-106 µm		Atomised	Spherical
	fine	<45 µm		Electrolytic	Irregular

The additive particle size has a negligible effect on the sintered density at the slow heating rate. A high heating rate induces a slight increase in the density. This is shown in figure 4.13 and table 4.6.

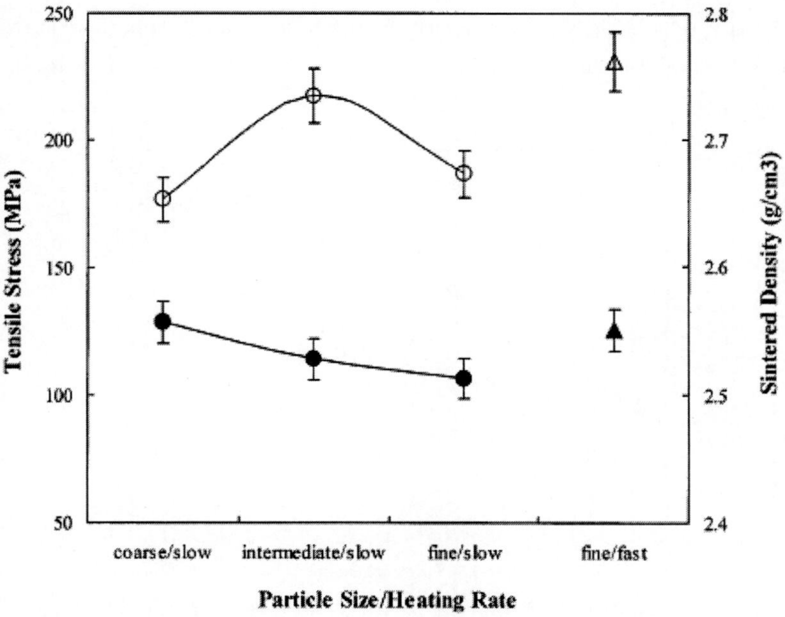

Figure 4.13: Effect of Particle Size and Heating Rate on the Sintered Density and Tensile Stress [13]

Table 4.6: Effect of Particle Size and Heating Rate on Density and Tensile Properties of Al-Cu Alloy [13]

Particle Size	**Heating Rate**	**Green Density (g/cm^3)**	**Sintered Density (g/cm^3)**	**Densification**	**Proof Stress (MPa)**	**Tensile Stress (MPa)**	**Elongation (%)**
Coarse	Slow	2.64	2.56	-0.51	123	177	3.8
Intermediate	Slow	2.62	2.53	-0.46	127	217	7.3
Fine	Slow	2.61	2.51	-0.49	118	187	4.8
Fine	Slow	2.61	2.55	-0.32	139	231	6.7

The tensile strength is maximised at either intermediate particle sizes or high heating rates. The proof stress, however, is not affected by particle size. This is evident from the load-displacement curves, shown in figure 4.14.

The work hardening rates in the three samples are similar and all the samples fail prematurely. It is therefore apparent that the high tensile strength at the intermediate particle size or at the high heating rate is due to improved ductility. As expected, the coarse additive particle sizes leave large residual pores behind. However, the inter-pore distance is also large and the matrix, particularly at the intermediate particle size, appears to be more intensely sintered. The material with the finest particle size and high heating rate has both fine secondary pores and a well sintered matrix.

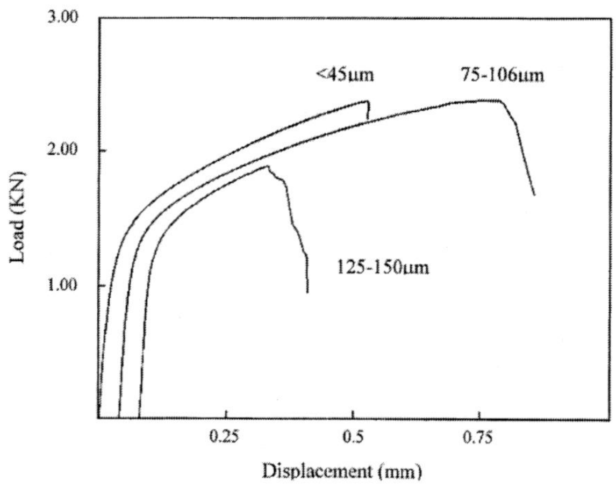

Figure 4.14: Typical Tensile Curves Showing Yield Strength and Ductility [13]

The tensile properties are correspondingly high. Metallographic examination of the gauge section of the fractured tensile specimens from the coarsest particle size indicated that the large secondary pores did not initiate fracture. Rather, cracking appeared to originate from poorly sintered regions.

The tensile properties of a sintered material are controlled by the density because pores reduce the load bearing area, concentrate the strain and create an incipient neck. Fine additive particle sizes can therefore be expected to be advantageous to sintered materials because they increase densification and leave smaller secondary pores in their place. However, it is apparent from Figures 4.13 and 4.14 that the properties in the Al-Cu system are not maximised at fine particle sizes but at intermediate sizes. It is therefore not simply the density of the compact and the size of the secondary pores that control the tensile properties. If bulk density was the major contributor to strength, then aluminium powder would not require much sintering because its relative softness allows very high green densities to be obtained by compaction alone. Green densities in excess of 90% are typical. Indeed, sintering of aluminium often causes expansion and results in negative densification values.

Coarse copper particles produce intensely sintered matrices. However, very coarse particles provide a highly localised and poorly distributed liquid. The matrix is only well sintered in parts and the properties are correspondingly low. Intermediate particle sizes are therefore preferred. Fracture of the coarse particle containing sample does not initiate at large secondary pores but at the poorly sintered inter-pore regions.

A high heating rate in transient systems also promotes liquid formation because it limits the time available for dissolution of the additive in the base prior to melting. This maximises the volume of the remaining liquid forming additive at the sintering temperature and hence the volume of liquid which ultimately forms. The effect of high heating rates on liquid formation is equivalent to coarse particles. The combination of fine particles and high heating rates produces an extensive and widely distributed liquid phase, a fine secondary pore size and high sintered densities. The resulting mechanical properties are therefore optimised.

Kempen et al. [15] investigated the mechanical properties of SLMed AlSi10Mg parts. He used a modified Concept Laser M1 SLM Machine was used to process the AlSi10Mg powder. This machine can handle reactive materials in a good controlled atmosphere. It is equipped with a 200W fibre laser and has a laser beam diameter of about 150 μm [16].

To determine the mechanical properties, tensile, hardness and Charpy impact tests were conducted. An Instron 4505 testing machine was used for tensile testing. Hardness was measured according to the hardness Vickers scale (0, 5 kg load).

In order to promote the optimum mechanical properties of AlSi10Mg parts produced by SLM, all test parts were produced using a set of parameters that result in a maximal density. The optimization of the parameters was done in a previous study [17].

Using a laser power of 200W, a scan speed of 1400mm/s and a spacing of 105μm between the scan tracks results in a relative density of at least to 98,5% The density can be further increased to 99,8% by re-melting every layer with the same parameters, but alternating directions over 90°. Mechanical properties obtained from the tensile tests done on parts produced in two different directions as shown in figure 4.15 and their observations are recorded in the table 4.7 in terms of the mechanical properties such as Hardness, strain, Ultimate tensile strength and young's modulus based on their respective building direction of the specimen.

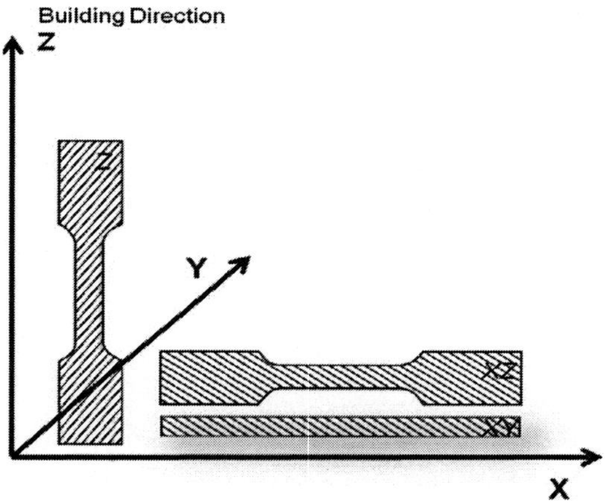

Figure 4.15: Building Directions of Test Samples [15]

Table 4.7: Mechanical Properties of SLM Built Parts and Cast + Aged Parts [15]

X± s	E(Gpa)	UTS(MPa)	ε break(%)	HV
XY direction	68±3	391±6	5,55±0,4	127
Z direction		396±8	3,47 ±0,6	
conventional cast and aged	71	300-317	2,5-3,5	86
high pressure die casting F	71	300-350	3-5	95-105
high pressure die casting T6	71	330-365	3-5	130-133

The given values represent the mean values for 2 or 3 specimens with 95% confidence intervals. As a reference, properties for conventional casted and high pressure die casted AlSi10Mg are also shown. High pressure die casting is considered to give the optimum properties amongst the casting processes. Stress-strain curve for each direction is shown in figure 4.16.

Observation shows that SLM AlSi10Mg parts have mechanical properties higher or at least comparable to the casted AlSi10Mg material. The Vickers hardness of the as built SLM parts is much higher (almost 30Hv) than the hardness of the high pressure die casted (HPDC) AlSi10Mg in the as-cast condition and almost as high as the HPDC AlSi10Mg in the aged condition. The ultimate tensile strength of the as built AlSi10Mg SLM parts is always higher than those of the HPDC in both conditions.

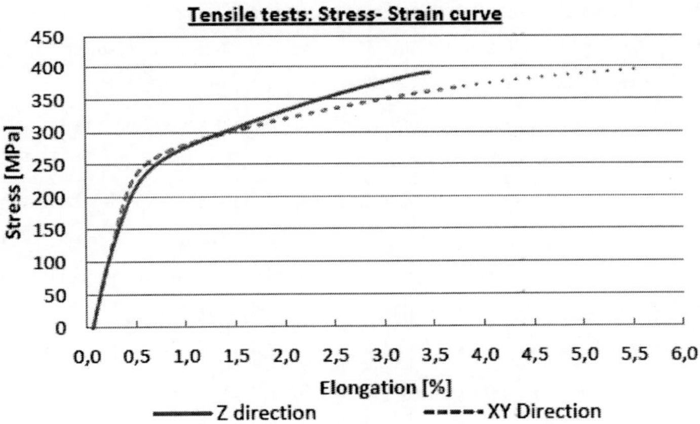

Figure 4.16: Stress-Strain Curve of SLMed-parts Produced in different Build Directions [15]

The elongation of the as-built AlSi10Mg parts in the Z-direction is comparable to the HPDC parts, while the elongation for parts built in XY – direction is almost 2% higher. Although the strength and hardness is higher, the Charpy impact energy as described in table 4.8 for the as-built SLM samples is still superior to that of the conventionally casted AlSi10Mg material.

Table 4.8: Charpy Impact Testing Values [15]

X±s	Impact energy (J)
XY direction	3,94± 0,5
Z direction	3,69 ±0,48
As cast	2,5 -3,0

Upon observation of the fracture surface of a tensile sample, it can be seen that larges pores near the border of the sample initiate the fracture. These 'borderline pores' are formed at the beginning/end of a scan vector. Borderline pores are more numerous in parts produced in the Z direction, compared to parts produced in the XY direction. These pores are the largest defects present in the part. At a high stress level (i.e. 395MPa for both testing directions), they will become the critical defects which initiate inhomogeneous deformation. Because of their location close to the sample border, they lack space for extensive deformation and easily cause complete fracture of the sample. Due to the faster strengthening of the Z direction, this high stress level is reached sooner and as a result the elongation in tensile test is lower: 3, 47% compared to 5, 55% for XY oriented samples.

4.6. Tribological behaviour of SLMed Parts

Prashanth, K.G., et al studied the tribological and corrosion properties of Al–12Si produced by selective laser melting by fabricating Cylindrical rods of 10 mm diameter and 15 mm length for wear experiments and square plates of 3 mm length, 3 mm width, and 2 mm thickness for corrosion experiments with nominal composition of Al–12Si (wt.%) have been produced by SLM. Heat-treatment of the SLM samples was carried out at 473, 573, 623, 673, and 723 K for 6 h under argon atmosphere. For comparison purposes, cylindrical Al–12Si rods and bars were also prepared by graphite mould casting. Sliding wear tests were carried out according to the ASTM G 99-05 standard at room temperature and in ambient atmospheric conditions using a pin-on-disc test device. A disc of 45 mm diameter and 13 mm thickness made of hard-faced stainless steel is used against the Al–12Si flat head pins of 9 mm diameter and 12 mm height. The tests were performed at constant load (10 N) with a sliding speed of 1 m/s for 30 min.

The wear rate was evaluated from the formula

$$Q_s = \frac{Vs}{Ls} \qquad (4.4)$$

L_s is sliding distance, V_s is the sliding volume loss and the subscript's' denotes sliding. Then the volume loss can be calculated using archard equation,

$$V_s = \frac{kWLs}{H} \qquad (4.5)$$

Where k is wear coefficient, W is applied load and H denotes hardness of the material. The volume loss was calculated from the wear loss determined by measuring the weight of the flat head pins before and after the tests.

The sliding distance is given by

$$L_s = 2\Pi r_s v_s t_s \qquad (4.6)$$

where r_s is the radius of the wear track (22.5 mm), v_s is the speed expressed in rounds per minutes (450 rpm), and t_s is the time (30 min).

For the immersion corrosion experiments, the square shaped plates were polished using SiC paper from 400 down to 4000 grit and subsequently polished using 3 μm and 0.25 μm diamond suspensions. The samples were cleaned with ethanol and the initial weight as well as the dimensions of the samples was measured. The samples were then immersed in 0.01 M, 0.1 M, and 1M HNO_3 solutions. The samples were removed from the acidic solution every 24h, subsequently rinsed with distilled water, dried in hot air and then weighed, and re-immersed in the acidic solution again for additional 24h. The corrosion products were not removed intentionally at any of the intermediate stages. They were removed only after the last measurement for microscopic investigations.

The wear rate, Vickers hardness, and size of the Si particles of the Al–12Si samples prepared by SLM are shown in figure 4.17 as a function of the annealing temperature along with the corresponding values of the as-cast Al–12Si material. The wear rate is at the minimum for the as-prepared SLM sample (300 K) and then it increases with increasing annealing temperature, following an exponential form very similar to the behaviour shown by the size of the Si particles in the same temperature range.

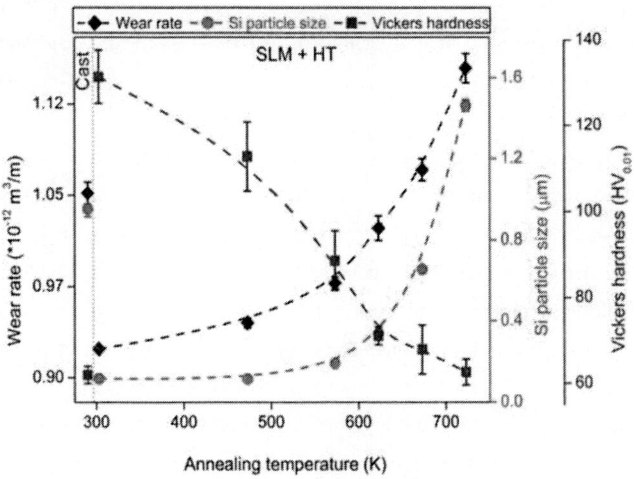

Figure 4.17: Sliding Wear Rate (υ), Average Size of the Si Particles (λ) and the Vickers Hardness (ν) for the Al–12Si Cast, as-prepared SLM (300 K) and SLM Samples Annealed at different Temperature [18]

References

[1] Polmear IJ. *Light alloys: metallurgy of the light metals.* 3rd ed. London: Arnold; 1995.

[2] Cam G, Kodak M. Progress in joining of advanced materials. *Int Mater Rev.,* 1998; 43:1–44.

[3] Mondolfo LF. Aluminium alloys: structure and properties. London: Butterworth's; 1976.

[4] http://www.aalco.co.uk/datasheets/Aluminium-Alloy_Introduction-to-Aluminium-and-its-alloys_9.ashx

[5] Ghosh, S.K., Bandyopadhyay, K., & Saha, P, "Development of an in-situ multi-component reinforced Al-based metal matrix composite by direct metal laser sintering technique — Optimization of process parameters", Materials Characterization, 2014, 93, pg.68–78.

[6] Ross PJ, "Taguchi techniques for quality engineering", New York: McGraw-Hill, 1996.

[7] Phadke MS, "Quality engineering using robust design", New Jersey: Prentice Hall: Englewood Cliffs, 1989.

[8] Yang WH, Tarng YS, "Design optimization of cutting parameters for turning operations based on the Taguchi method", J Mater Process Technol, 1998, 84, pg. 122–9.

[9] Kim KD, Choi DW, Choa YH, Kim HT, "Optimization of parameters for the synthesis of zinc oxide nano particles by Taguchi robust design method", *Colloids Surf A Physicochem Eng Asp* 2007, 311, pg. 170–3.

[10] Jiang QC, Ma BX, Wang HY, Wang Y, Dong YP, " Fabrication of steel matrix composites locally reinforced with in situ TiB2–TiC particulates using self-propagating high-temperature synthesis reaction of Al–Ti–B4C system during casting", *Compos Part A,* 2006, 37: 133–8.

[11] Ghosh SK, Saha P, Kishore S, "Influence of size and volume fraction of SiC particulates on properties of ex situ reinforced Al–4.5Cu–3Mg metal matrix composite prepared by direct metal laser sintering process", *Mater Sci Eng A* 2010, 527: 4694–pg. 701.

[12] Buchbinder D, Schleifenbaum H, Heidrich S, Meiners W, Bultmann J, "High power selective laser melting (HP SLM) of aluminium parts. *In: Physics procedia of the sixth international WLT conference on lasers in manufacturing",* vol. 12A, 2012, Munich ICM, Germany; p. 271–8.

[13] Lumley, R., & Schaffer, G. (1998). The effect of additive particle size on the mechanical properties of sintered aluminium-copper alloys. *Scripta Materialia, 39*(8), 1998, pg. 1089–1094.

[14] Buchbinder, D., Meiners, W., Wissenbach, K., Müller-Lohmeier, K., Brandl, E, "Rapid Manufacturing of Aluminum Parts for Serial Production via Selective Laser Melting (SLM)", In: *Aluminum Alloys, Ed. J. Hirsch [u.a.], Wiley,* Vol. 2, 2008, pg. 2394-2400.

[15] Kempen, K., Thijs, L., Van Humbeeck, J., & Kruth, J.-P, "Mechanical Properties of AlSi10Mg Produced by Selective Laser Melting", *Physics Procedia, 39,* 2012, pg. 439–446.

[16] Concept Laser GmbH/ M1, Web-Based Data, Concept Laser GmbH Co., Germany, http://www.concept-laser.de/, as on 04.05.2012.

[17] Kempen, K., Thijs, L., Yasa, E., Badrossamay, M., Verheecke, W., Kruth, J, "Microstructural analysis and process optimization for sleective laser melting of AlSi10Mg", Solid Freeform Fabrication Symposium Proceedings. Solid Freeform Fabrication Symposium. Austin, Texas, USA, 8-10 August 2011.

[18] Prashanth, K.G., Debalina, B., Wang, Z., Gostin, P.F., Gebert, A., Calin, M., Eckert, J, "Tribological and corrosion properties of Al–12Si produced by selective laser melting", *Journal of Materials Research,* 29(17), 2014, pg. 2044–2054.

Chapter V

Selective Laser Melting (SLM) of Magnesium (Mg) Alloys

CHAPTER V

5. Selective Laser Melting (SLM) of Magnesium (mg) Alloys

5.1. Introduction

Magnesium is the eighth-most-abundant element in the Earth's crust by mass and tied in seventh place with iron in molarity [1]. Magnesium is commercially produced by electrolysis of molten magnesium chloride (MgCl2), processed mainly from seawater and by the direct reduction of its compounds with suitable reducing agents. At one time, magnesium was used for photographic flash ribbon and powder, because in finely divided form it burns in air with an intense white light; it still finds application in explosive and pyrotechnic devices. Because of its low density (only two-thirds that of aluminium), it has found extensive use in the aerospace industry.

Magnesium is the easiest structural metal to machine and has often been used when a large number of machining operations are required. The advantages of magnesium are listed below [2, 3],

- Lowest density of all metallic materials.
- High specific strength.
- Good cast ability.
- Corrosion resistant.
- Good Machinability.
- Good impact resistance.
- Good damping property.

5.1.1. Application of Magnesium

Magnesium alloys have a number of applications: they are used for parts of aircraft, spacecraft, machinery, automobiles, and portable tools to electronic and biomedical field.

5.1.2. Structural Applications

Automotive, aerospace, industrial, and commercial applications are examples of structural applications.

Automotive: support brackets for brakes and clutch, housing for transmission.

Aerospace: landing wheels, helicopter rotor fittings, gearbox housings.

Industrial: high-speed operating machinery, such as textile machines.

Commercial: luggage, hand tools, computer housings, ladders.

5.1.3. Electronic Applications

Electronic applications include electronic packaging, hard drive arms, cell phone and portable media device housings. Magnesium alloys are being used instead of plastics due to their light weight, strength and durability. They also are relatively better for heat dissipation and protection against electromagnetic and radio frequency interference [4].

5.1.4. Medical Applications

Portable medical equipment and wheelchairs that require light material make good use of magnesium alloys. Also, cardiovascular stents and orthopaedic devices are potential applications of some magnesium alloys due to magnesium's biocompatibility and bio absorbability [5].

5.2. Densification Behaviour based on Laser Related Process Factors

Zhang et al. [6] studied the influence of laser energy densities on the densification of Mg-9%Al (Mg with irregular shape and Al with atomized spherical shape) alloy parts by setting the laser power and laser scan speed using a continuous wave Nd: YAG laser ranges between 34 μm and 75 μm. The SLM was processed under a closed environment filled with 7 mbar argon atmosphere with oxygen content less than 0.5%. The operating temperature of the internal chamber was fixed at 80°C. Each powder layer thickness 50 μm and a hatch spacing 80 μm were chosen for this study.

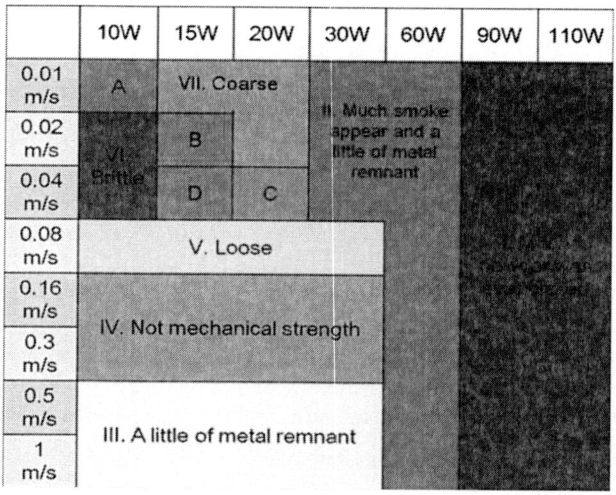

Figure 5.1: Effect of Laser Related Parameter Variation in Mg-Al Alloy [6]

Figure 5.1 shows the seven different processing scales based on the laser power and scanning speeds [6]. (I, II) High energy input zones: the delivered energy density of the laser beam is extremely high in this zone under a high laser power for all speeds used. During the experiment, much powder was evaporated with high energy density due to the low boiling point of Mg, and then oxidized to form MgO mote filled in the protect- gas chamber. Finally only the ultra-thin metal layer solidified on the substrate with irregular shape after each powder layer was sintered. (III, IV, V) Low energy input zones: a significantly low energy input results in a surface melting of particulate under a low laser power and a high scanning speed, and a poor bond neck between the particulates can be obtained.

The obtained samples displayed a structure of powder stacking without any mechanical strength under the laser parameters as shown in zones III and IV. The samples in zone V presented a loose metal structure. (A, B, C, D, VI, VII) Forming zones: laser sintering at a low laser powers (10–20 W) and a low scanning speed (0.01–0.04 m/s) can generate a sufficient density of energy input. Meanwhile, several brittle samples with stratification caused by limited melting are obtained in zone VI. The coarse surface of samples exhibited an agglomeration phenomenon when using laser parameters in zone VII. It can be explained that the relatively high laser power and a long liquid lifetime lead to a strongly surface tension ("balling" effect) [7]. Finally, four samples (A, B, C, and D) processed by laser energy density 20 J/mm2, 15 J/mm^2, 10 J/mm^2, and 7.5 J/mm^2.

It was observed that the increase in laser energy density from 7.5 J/mm^2 to 15 J/mm^2 resulted in improved relative density of the melted samples from 74.5% to 82%. At lower laser energy densities (higher speeds), there is a significant porosity caused by incomplete melting leading to formation of discontinuous scanning tracks with large sized balls i.e. balling. With increase in laser energy density, better melting of the powders was achieved enabling more liquid phase to flow and infiltrate the voids between the particles and consequently, the pores dispersed and diminished to form a relatively smooth surface with increased density. Maximum densification was achieved under a lower laser power input of 15 W and a scanning speed of 0.02 m/s. However, further increment in the laser energy density to 20 J/mm^2 led to a reduction in relative density to 76.1%. Though a sufficient amount of liquid phase was generated at higher laser powers, reduced scan speed causes reduction in the surface energy of the liquid at short length scale causing spheroidisation and breakdown in the melt pool [8]. Further. They concluded that a critical scanning speed of 0.02 m/s was required to ensure complete melting of the powders without evaporation and a sound part was not obtained above a speed of 0.04 m/s for all the laser powers used.

Liu, S., & Guo, H et al. [9] investigated the influence of hot isostatic Pressing on mechanical properties of SLMed magnesium alloy with the particle size of spherical AZ61 powder is 30-70μm. The samples were fabricated by using the FORWEDO LM-120 under argon protection. Considering the forming efficiency and quality of process related factors such as the laser power with 150W, the scanning speed 400mm /s, the hatch spacing 0.06mm, the layer thickness is 0.04mm, the laser spot diameter is 0.07mm, and the scanning strategy is zigzag with an angle of 90° between adjacent layers. The hot isostatic pressure(HIP) test was performed at 450°, 103MPa, 3hrs.The surface topography after polishing with a reagent of 50 ml ethyl alcohol, 2ml HNO_3, and phase analysis were determined by optical microscopy (Leica DM4M) and X-ray diffraction (MXP21VAHF XRD), respectively. The microstructure fracture morphologies were studied by SEM (JSM-6701F) and the figure 5.2 shows the pores of SLM and SLM + HIP specimen.

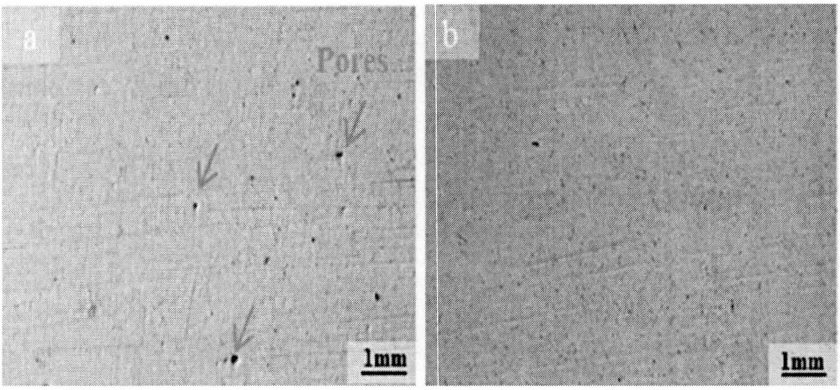

Figure 5.2: Cross-section of: (a) the SLM Specimen; (b) the SLM + HIP [7, 9]

It can be seen from figure 5.2 [9] that the pores are significantly reduced after HIP. According to Image pro software statistics, the pore area fraction of SLM specimen is 0.8%, and the average pore size is 43 ± 27 μm. Most of the pores are closed after HIP, and the density is close to 100%.

5.3. Defects in SLM of Magnesium Alloys

Zhang et al. [6] studied the effects of scanning speed and laser beam powers on the porosity of the melted samples are shown in figure 5.3. It can be observed that the porosity decreases with raising energy density from parameter D to B. It can be found that the porosity reduces from 25.5% to 18% when the energy density 7.5 J/mm^2 increases to 15 J/mm^2. In contrast, the porosity increases from 18% to 23.9% when the energy density increases from 15 J/mm^2 to 20 J/mm^2.

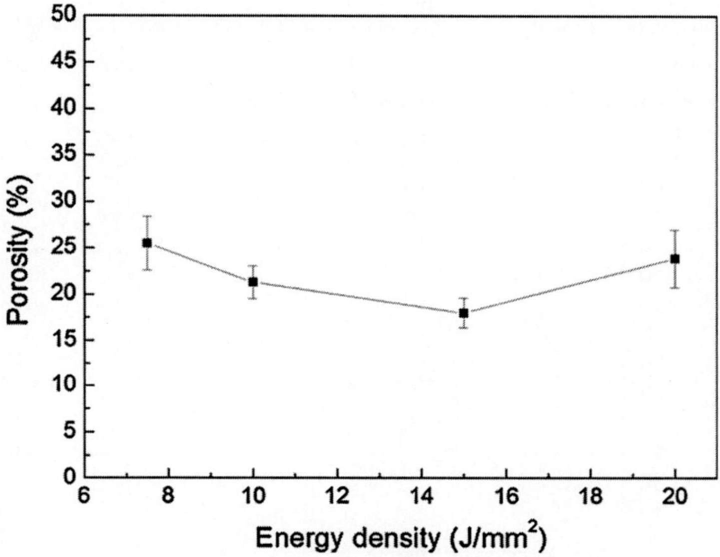

Figure 5.3: Effect of Energy Density on Porosity for Melted Samples [6]

As a result, high energy density input could improve liquid– solid wetting characteristics, and better melting fluidity to prepare porous less samples. Base on this theory, it can be confirmed that low porosity is associated to relatively high energy density, whereas the result of experiment showed that the evaporation of metal from the pool increases the porosity of samples when high energy input intensity was employed (figure 5.1, Zone II), due to the low boiling point and low laser energy absorption rate of Mg–Al powder. In this study, the lowest porosity is about 18% by means of the parameter B (15 W, 0.02 m/s).

5.4. Mechanical Properties of SLMed Part or Component

5.4.1. Hardness

Zhang et al. [6] studied the effects of scanning speed and laser beam power on hardness of the melted samples is shown in figure 5.4. It can be observed that there is a minor difference of the melted samples hardness with different laser beam parameters. It can be observed that the tolerance of sample D is much larger than the others from figure 5.4, which can be explained by a localized zone of un-molten powder.

Based on References [10, 11], the hardness of pure Mg and Al_2O_3 is 30 HV25–40 HV25, and 140 HV25–150 HV25, respectively. It can be concluded that the Mg–Al powders is un-molten with the parameter of D (15 W, 0.04 m/s). Furthermore, Mg and Al powder was still segregated after sintering (Al_2O_3 can be seen as little amount of remainder Al was completely oxidized). Therefore, the hardness of other samples under the parameters (A, B, C) are similar to the hardness of the original cast Mg–Al alloys (AZ91 or AZ31) 60 HV25 [12].

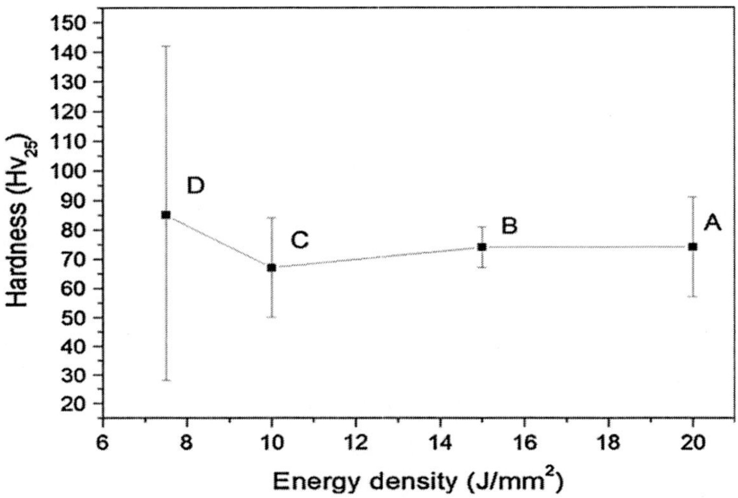

Figure 5.4: Effect of Energy Density on Hardness for Melted Samples [6]

Liu, S., & Guo, H [9] as discussed in section 5.2 also investigates the mechanical properties of SLMed and SLMed + HIP specimen. The microhardness and in-situ tensile stress distributions were examined by LeicaVMHT30M (loading force of 25gF, holding time of 10sec) and WDW-200S DIC device, respectively. The tensile test was carried on by the CMT4105 material testing machine at room temperature. Figure 5.5(a)shows the intensity of the peaks (4 1 1) and (2 4 2) of the b- $Mg_{17}Al_{12}$ phase weakened significantly after HIP, indicating that most of the b phase was dissolved into the matrix after HIP, and the peak intensity of Mg was enhanced, the crystallinity of Mg is improved.

At the same time, it is clear that the microhardness of the SLM specimen is about 98.9 ± 5.9Hv, and the microhardness after HIP is reduced to 87.4 ± 3.9Hv, which is higher than that of the forged specimens of 63.5 ± 3.4Hv. Combining figure 5.5 (c), (d), and the brittle-hard b-$Mg_{17}Al_{12}$ changed from continuous precipitation along the grain boundary to discontinuous precipitation after HIP, so the hardness decreases slightly. The grain size of the SLM specimen is much smaller (2.2 ± 0.3 lm) than that of SLM + HIP specimen (23.9 ± 6.0 lm) by line intercept method because of the exorbitant temperature or the over holding time.

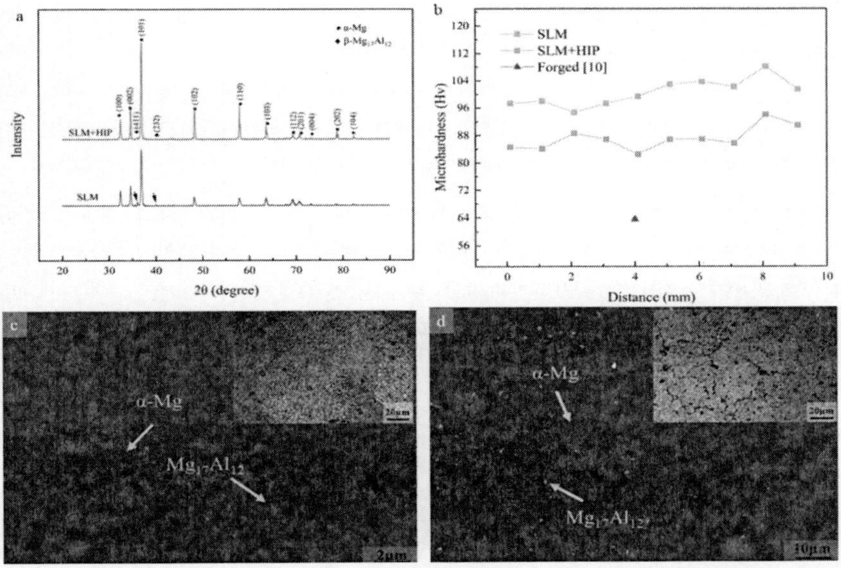

Figure 5.5: (a) XRD Patterns of SLMed and SLMed+ HIP Specimen, (b) Hardness of SLMed, SLMed + HIP and Forged Specimen, (c) SEM Image of SLMed (d) SEM Image of SLMed+ HIP Specimen [9]

Dutta Majumdar, J et al. [13] investigated the hardness of MEZ and LSM MEZ alloy. Laser Surface Melting was carried out using a 10 kW continuous wave CO_2 laser (Model: Rofin Sinar, RS 10000) having a beam diameter of 4mm with Ar as shrouding environment (at a flow rate of 6 l/min) to avoid oxidation during lasing.

Table 5.1: Laser Parameters Employed for LSM of MEZ [13]

Sample No.	Laser power(kW)	Scan rate(mm/min)
1.	1.5	100
2.	1.5	150
3.	1.5	200
4.	2.0	150
5.	2.0	200
6.	2.0	250
7.	2.5	300
8.	2.5	150
9.	2.5	200
10.	2.5	250
11.	2.5	300
12.	3.0	200
13.	3.0	250
14.	3.0	300

The specimens were mounted on a water chilled copper block placed on a CNC controlled stage that was moved at a speed of 100–400 mm/min. The laser focus rested 30mm above the sample surface. The process parameters variable for the present studies were laser power and scan speed as shown in table 5.2. Adequate number of trials was conducted to correlate the process parameters with the microstructure and optimize the laser processing parameters.

Figure 5.6 represents (Microhardness profile as a function of depth from the surface on the cross-sectional plane of (1) as-received MEZ and laser surface remelted MEZ; (2) with laser power = 2 kW, scan speed = 200 mm/min; (3) laser power = 2.0 kW, scan speed = 300 mm/min; (4) laser power = 1.5 kW, scanspeed = 200 mm/min, respectively).

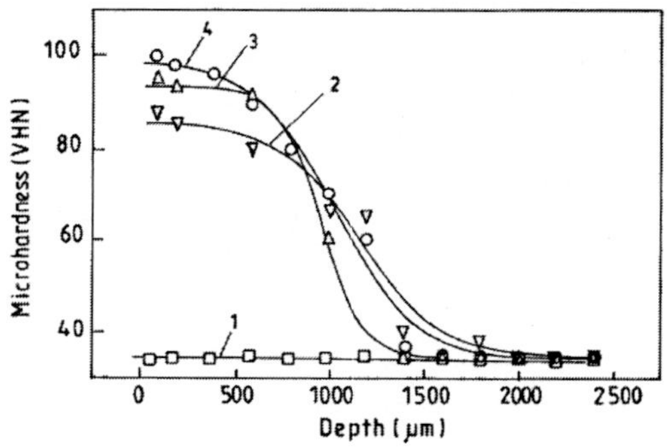

Figure 5.6: Microhardness for MEZ and LSM MEZ with Process Factors [13]

The microhardness profiles of the laser surface melted MEZ specimens (lased under a few optimum distance from the surface measured on the cross-sectional plane in the laser surface melted zone. It is apparent that the microhardness of the melted zone has significantly increased by 2 to 4 times (85–100 VHN) as compared to that of the substrate (35 VHN).

The microhardness of the remelted zone is maximum, near the surface of melted zone and decreases gradually towards the interface of the melted zone. This improvement in microhardness due to laser surface melting may be attributed to grain refinement and solid solution hardening. Furthermore, the average microhardness in the laser surface melted zone seems to vary with the laser parameters (laser power and scan speed). It is apparent from figure 5.6 that the maximum microhardness is achieved following laser surface melting, with a power of 1.5kW and a scan speed of 200 mm/min. It may further be noted that lasing with a higher power produces coarser microstructure (due to slower cooling and/or heat conducted from the overlapping tracks), and hence lowers microhardness (plot 3 versus plot 1). Similarly, lasing with a higher scan speed improves the microhardness (plot 1 versus plot 2) because of higher cooling rate associated with the higher scan speed. Hence, it may be concluded that the significant enhancement in microhardness and refined microstructure produced by laser surface melting may be effective in imparting a better resistance to wear and erosion in MEZ components.

Cui, Z., Shi, H et al. [14] and his teammates in investigated Laser surface melting AZ31B magnesium alloy with liquid nitrogen-assisted cooling with a 3 mm thick AZ31B Mg alloy sheet with 3.22% Al, 1.15% Zn, 0.400% Mn, 0.0133% Si, 0.0019% Fe (mass fraction), and balanced Mg was used in this work. The material was cut into specimens with dimensions of 100 mm×100 mm×3 mm for laser treatment. LSM was carried out using a pulsed Nd:YAG laser with a wavelength of 1064 nm. The laser processing parameters, optimized from adequate number of preliminary experiments during the present efforts, are listed in table 5.3. Prior to laser treatment, the specimens were ground with an 800 SiC grit so as to eliminate surface oxides and obtain cleaned surfaces, then rinsed in ethanol, and finally dried in air.

Table 5.2: Parameters of LSM [14]

Voltage U/V	**Laser pulse time, t_p/ms**	**Repetition rate, f /Hz**	**Defocusing amount, (Δz) /nm**	**Diameter, d/mm**	**Scanning rate, ν mm/s**
62	3	30	+3	2	7

The plate was immersed in a steel container with liquid nitrogen (−196 °C) resulting in about 2 mm-thick deep cryogenic liquid layer over the target to reach rapidly cooling rates and avoid surface oxides, as shown in figure 5.7.

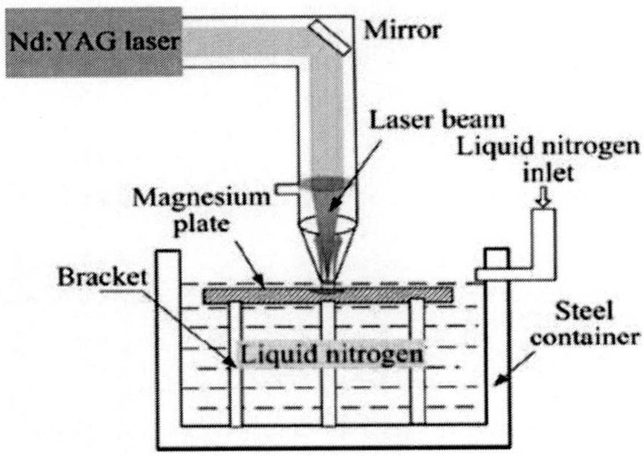

Figure 5.7: Schematic Layout of LSM in Liquid Nitrogen [14]

An overlap ratio of 30% between the laser-melted tracks was selected during laser melting. In order to contrast test results, another plate was melted with the same parameters by laser and with 99.99% high purity Ar gas (gas flow of 5 L/min) protection environment at room temperature and the specimen with the presence of liquid nitrogen is compared with Ar at room temperature. Hardness is compared with the specimen with Ar atmosphere and in the presence of liquid nitrogen was compared.

The vibrations of microhardness along depth of laser melted layers are depicted in figure 5.8. It is apparent that the microhardness of laser melted layer in liquid nitrogen has significantly increased to HV 90–148 as compared to that of laser melted layer in Ar gas protection (HV 65–105).

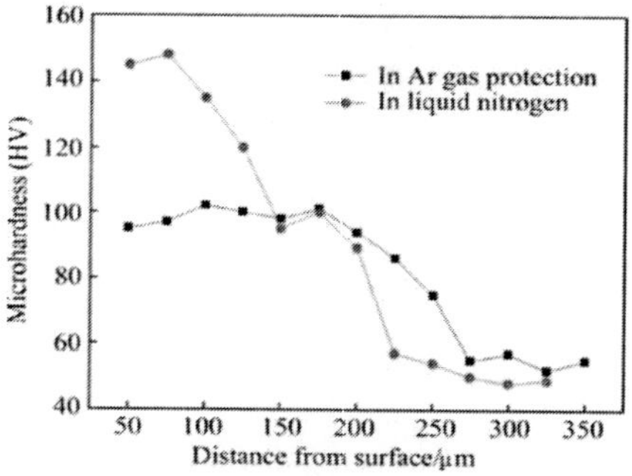

Figure 5.8: Microhardness along the Depth of Laser Surface Melting (LSM) [14]

The maximum micro- hardness of the melted layer in liquid nitrogen appears near the surface and the whole change trend is larger than the latter. For the latter, the increase of microhardness is mainly due to the fine dendrites microstructure, the re-dissolution and re-distribution of β-phase and dislocation. According to Hall–Petch equation:

$$\sigma = \sigma_0 + kd^{-1/2} \quad (5.1)$$

Where, σ is the yield stress, σ_0 is the yield stress of a single crystal, k is a constant and d is the grain size.

The improvement factors of microhardness of the melted layer in liquid nitrogen may be different with the above. Some research results indicated that the mechanical properties of nano crystalline materials are usually much higher than its polycrystalline. Hall–Petch equation 5.1 of the relationship between strength and grain size does not continue to nanoscale. This is due to that Hall–Petch formula derived from the reinforcement of dislocation. But when the grain sizes reach nanoscale, the dislocation may be less or even does not exist. Therefore, the enhancement of microhardness of the melted layer in liquid nitrogen may be attributed to the changes of β-phase morphology and distribution. In addition, local amorphous structure can also help to improve the microhardness of the melted layer.

5.4.2. Tensile Strength

Figure 5.9(a), (b) compares the mechanical properties of the SLM specimen and SLM + HIP specimen. The ultimate tensile strength (UTS) and Yield tensile strength (YTS) of SLM and SLM + HIP are 287 MPa, 233 MPa and 274 MPa, 126 MPa, respectively. The elongation is greatly improved compared to SLM, from 3.1% to 8.2% without reducing the UTS. According to the Hall-Petch formula [10, 15] the yield strength of HIP specimen decreases with increasing grain size of 23.9 ± 6.0 μm. However, the increase in density of HIP specimen allows the strength to be maintained at a higher level.

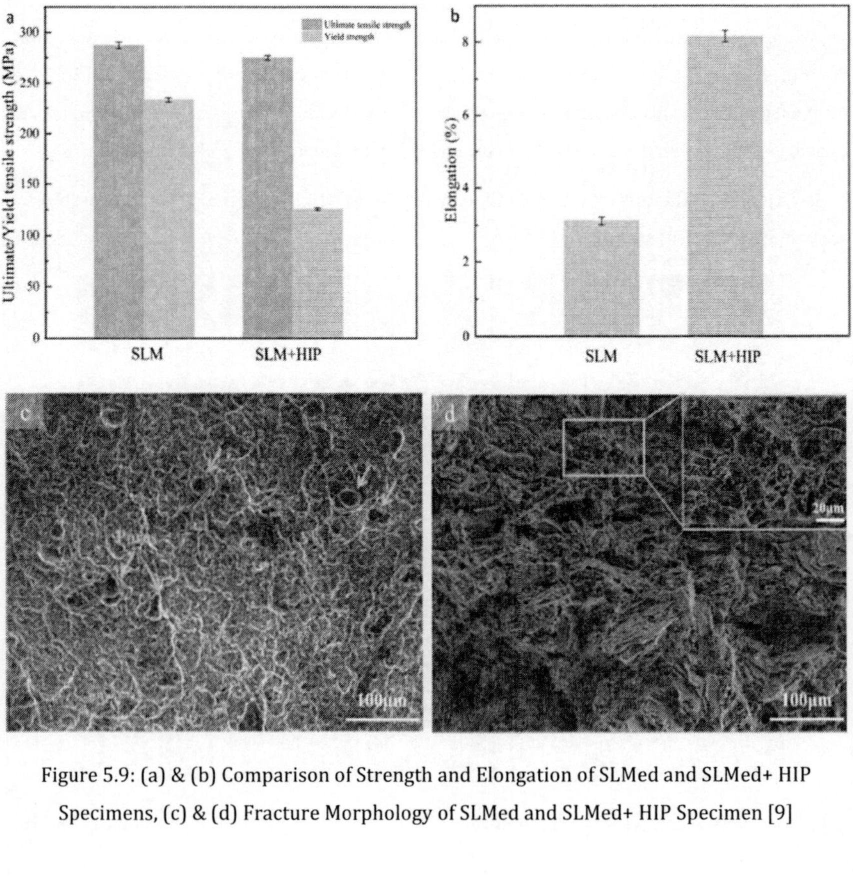

Figure 5.9: (a) & (b) Comparison of Strength and Elongation of SLMed and SLMed+ HIP Specimens, (c) & (d) Fracture Morphology of SLMed and SLMed+ HIP Specimen [9]

On the other hand, according to the solid solution strengthening effect [15], the solid solution atom Al concentration increases after the $Mg_{17}Al_{12}$ solid solution, which plays a strengthening role, so the material reaches a high strength close to that of the SLM specimens. The plasticity improvement may be attributed to the following two points: First, densification increases with HIP treatment, reducing defects such as pores that cause fracture. Second, due to the high dislocation density and stress concentration at the interface between brittle-hard b and the a-Mg matrix, so that the $Mg_{17}Al_{12}$ on the grain boundary provide additional sites for crack nucleation, a crack source will be generated at the interface during tensile. It is worth noting that $Mg_{17}Al_{12}$ at the grain boundary is dissolved after the HIP. Hence, the combination of the reduction of the crack source and the weakening of the pinning effect of the $Mg_{17}Al_{12}$ at the grain boundaries leads to an increase in plasticity.

The fracture morphology of SLM specimen and SLM + HIP specimen are shown in figure 5.9 (c), (d). The fracture of SLM is composed of small, shallow dimples with some obvious pores. The pores in the fracture morphology of SLM + HIP were significantly reduced, and mainly composed of large and deep dimples and cleavage steps, the co-occurrence of lot of dimples and cleavage steps revealed the mixture fracture mode of ductile fracture and brittle fracture.

Owing to the different total tensile time, typical stress distribution diagrams were selected according to different stages.

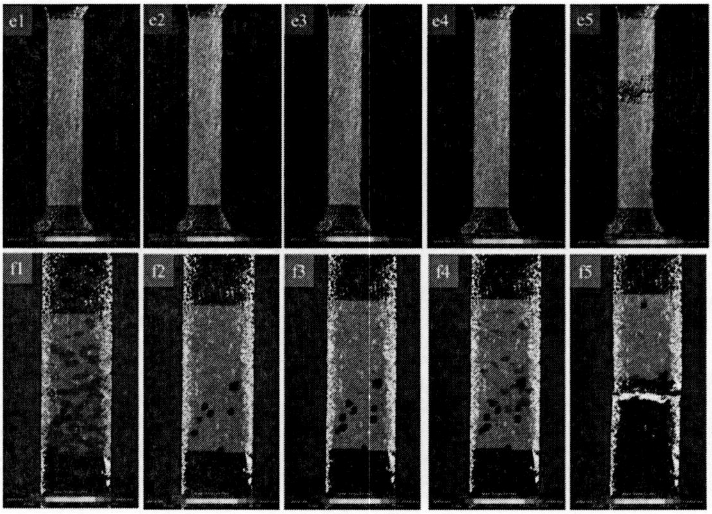

Figure 5.10: (e1–e5), (f1–f5), Digital Image Correlation (DIC) Diagram of SLM and SLM + HIP Specimens (SLM + HIP Specimen [9]

It can be seen from the Digital Image Correlation (DIC) diagram of SLM and SLM + HIP specimens in situ tensile test in figure 5.10(e1–e5), (f1–f5), SLM + HIP specimen experienced local stress concentration due to the large grain size during the entire tensile process, the local maximum strain value reached 12.4%, which was in contrast to the uniform stress distribution of SLM specimen with the highest strain value of 1.02%. It has been pointed out that the initial grain size should be reduced. When they were deformed at room temperature, because large-grained alloy twined at higher strains, leading to fracture whereas smaller grain size led to more homogeneous distribution of stress concentration in the grain boundaries [16]. Without sacrificing UTS, the elongation has been increased by 160% through HIP. The next step should be to control the grain size and increase the yield strength by shortening the holding time or reducing the temperature.

5.5. Tribological Behaviour of SLMed Part or Component

5.5.1. Corrosion

Dutta Majumdar, J et al. [13] as discussed in section 5.4.1with various processing factors investigated the effect of Laser Surface Melting(LSM) on corrosion and wear resistance of a commercial magnesium alloy using Magnesium with rare earth elements (MEZ - Zn 0.5%, Mn 0.1%, Zr 0.1%, rare earth elements 2%, Mg remaining percentage)to improve its corrosion property. Pitting corrosion resistance of the samples before and after laser surface melting was studied in a 3.56 wt. % NaCl (pH = 5) solution for a period of 0.5–72 hours by standard immersion test (ASTM G-46) [17]. The specimen was carefully polished prior to immersion for a predetermined period of time. After immersion, the specimen was ultrasonically cleaned and observed under microscope. The pit density (area fraction of pits as per standard point counting method) [18] is a function of time (t) was determined by careful optical microscopy.

The pitting corrosion behaviour of the laser surface melted specimen was also compared with that of untreated one by calculating the corrosion rate derived from the potentiodynamic polarization study (in a 3.56 wt.% NaCl solution in air) [19]. It is evident form figure 5.11 i.e. formation of pit formation (in terms of area fraction of pits as a function of immersion time) in (1) as-received and (2) laser surface melted (with a power of 2 kW, scan speed of 200 mm/min) MEZ samples followed by immersion test in a 3.56 wt. % NaCl solution) that both the extent and rate of pitting are significantly reduced following laser surface melting. In as-received MEZ, pit formation starts even after 30 min of immersion time.

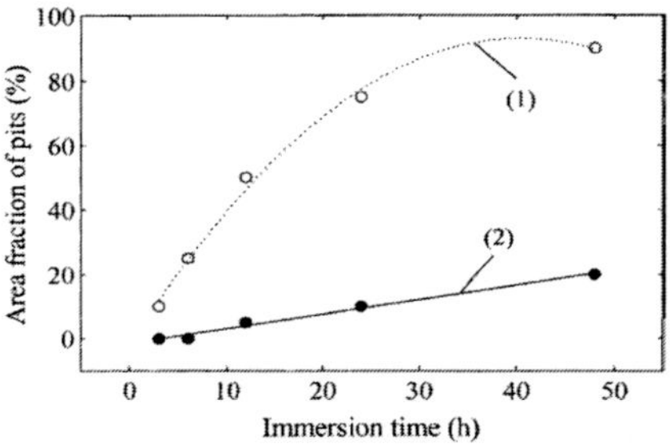

Figure 5.11: Kinetics of Pit Formation [13]

Furthermore, the rate of pitting is initially faster and decreases after 30 h of immersion time in as-received MEZ. After prolonged exposure, the pits grow in lateral direction and get interconnected, leading to almost constant number of pit density. On the other hand, visible pits are observed in laser surface melted MEZ only after 12 hours of exposure and the kinetics of pitting is much slower than that of the as-received MEZ. The pits remain isolated (not interconnected) even after 72 h of exposure. There was a lower area fraction of pits because they were smaller (isolated).

The pits form due to electrochemical degradation of material. Since this process involves transport of charged species, corrosion studies were performed with as-received and laser surface melted MEZ samples under applied potential using both potentiostatic and potentiodynamic conditions. During the potentiostatic immersion test, the specimen was polarized at a potential nobler than the open circuit potential (summarized in table 5.4) for a predetermined period of time, and the corrosion rate was determined from the weight loss of the specimen as a function of applied potential. Figure 5.12 shows the variation of weight loss of as-received and laser surface melted (lased with a power of 2 kW and a scan speed of 200 mm/min) MEZ samples, as a function of applied potential during isochronal (30 min) immersion in a 3.56 wt.% NaCl solution.

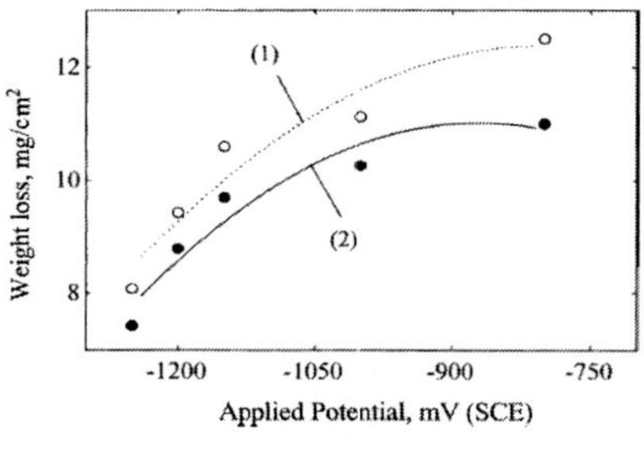

Figure 5.12: Corrosion Loss of Specimen [13]

It is apparent that weight loss due to corrosion increases with increase in applied potential. Furthermore, weight loss due to corrosion is higher in as-received MEZ as compared to that of the laser surface melted one.

Figure 5.13 (a) and (b) (a) as-received and (b) laser surface melted (with a power of 2 kW and a scan speed of 200 mm/min) MEZ samples) shows the Tafels plot derived from the potentiodynamic polarization studies[19].

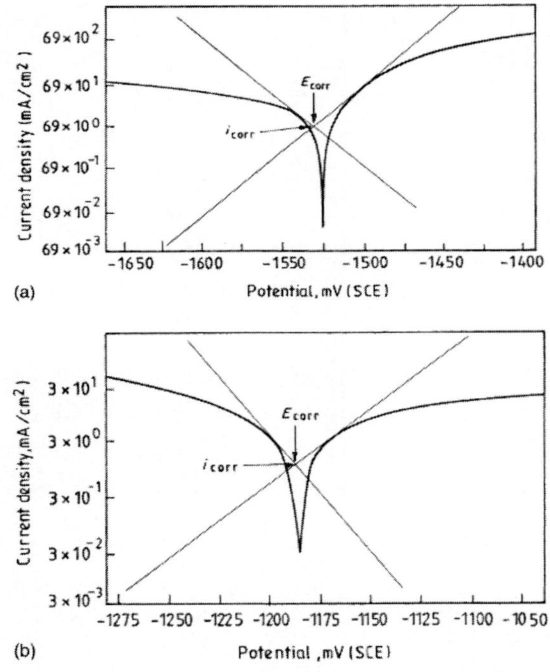

Figure 5.13: a) & b) Variation of Current Density with Potential [13]

The corrosion current (i_{corr}) and corrosion potential (E_{corr}) values for the respective samples, are obtained from the intersection of cathodic and anodic Tafel plots. Accordingly, the corrosion rate is determined using the relationship:

$$\text{Corrosion rate (mpy)} = (C \times W \times i_{corr}) / \rho \qquad (5.2)$$

Where, C is the conversion factor, W is the equivalent weight of test specimen (g), ρ is the density of the material (mg/m^3), and icorr is the current density (mA/m^2). Polarization resistance, which is an indirect measure of corrosion resistance, has subsequently been calculated using the relation [20]:

Polarization resistance $\frac{dE}{di}$ = (β_a × β_c) / (2.3 i_{corr}× (β_a + β_c) (5.3)

Where, β_a and β_c are the slopes of tangents drawn on the respective anodic and cathodic polarization plots. The calculated values of corrosion current, corrosion voltage, and corrosion rate are presented in Table 5.3.

Table 5.3: Corrosion Rate and Polarization Resistance of MEZ in a 3.56 wt.% NaCl Solution[13]

Sample	Open circuit potential (mV) (SCE)	E_{corr} (mV) (SCE)	i_{corr} (mA/cm^2)	Corrosion rate(mpy)	Polarization resistance (mΩcm^2)
MEZ	-1525	-1530	69	6.12	15.07
Remelted	-1185	-1190	1.5	0.133	899.9

A detailed analysis of the corrosion data in table 5.3, reveals that laser surface melting has significantly reduced the corrosion rate to 0.133 mpy as compared to 6.12 mpy of the substrate in a 3.56 wt.% NaCl solution. Furthermore, the polarization resistance, which is a measure of corrosion resistance (dE/di), was also found to increase significantly as a result of laser surface melting.

Cui, Z., Shi, H [21] and his teammates in investigated Laser surface melting AZ31B magnesium alloy with liquid nitrogen-assisted cooling with a 3 mm thick AZ31B Mg alloy sheet with 3.22% Al, 1.15% Zn, 0.400% Mn, 0.0133% Si, 0.0019% Fe (mass fraction), and balanced Mg was used in this work as discussed in section 5.5.1 as same process parameters, the corrosion resistance has been compared. The potentiodynamic polarization curves of laser melted layers and as-received AZ31B Mg alloy are shown in figure 5.14.

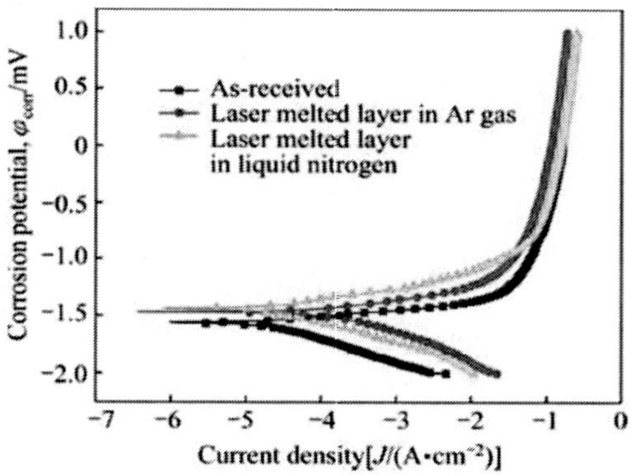

Figure 5.14: Potentiodynamic Polarization Curves of Laser Melted Layer and As-received AZ31B Mg Alloy [21]

Table 5.4 shows the corrosion potential, mass loss and corrosion rate of as-received AZ31B and laser melted layer. The corrosion potential of laser melted layers in liquid nitrogen is −1439 mV, which is 26 mV and 124 mV higher than those of the layers at room temperature and as-received Mg alloy substrate, respectively.

Table 5.4: Electrochemical Test Values for Corrosion [21]

Sample	φ/mV	Pre – corrosive mass, m_1/g	Post – corrosive mass, m_2 /g	Mass loss /g	Average corrosion rate /(gm $cm^{-2}.h^{-1}$)
Base metal	-1563	3.943	3.962	0.019	0.0684
In Ar gas	-1465	3.736	3.745	0.009	0.0326
In liquid nitrogen	-1439	4.445	4.448	0.003	0.0108

The corrosion rate of melted layers in liquid nitrogen is 0.0108 g/ ($cm^{-2}h^{-1}$), which is about 1/3 of that of the layers at room temperature. Both of the melted layers have an improved corrosion rate compared with the as-received substrate. Therefore, we can conclude that the corrosion resistance of samples in deep cryogenic treatment is improved compared with that of the samples in Ar gas protection.

Majumdar et al [13] thought that the Mg corrosion mechanism consisted of two distinct stages: one is initiation of corrosion dominated by mainly galvanic corrosion and the other is propagation of corrosion dominated by pitting corrosion. There is always a surface film on Mg and its alloys even after polishing, and the film mainly consists of magnesium oxides and hydroxides, which has been confirmed by XRD and XPS [22].

5.5.2. Wear

Dutta Majumdar, J [13] discussed with same process discussed in section 5.5.1 investigated the wear loss (in terms of reduction in thickness) of as-received and laser surface melted MEZ corresponding to the maximum average microhardness value (lased with a power of 1.5kW and a scan speed of 200 mm/min) was measured at regular intervals by a Pin-on-Disc wear testing machine against a hardened steel disc with a 3 kg applied load and 300 rpm wheel speed for a period between 1 min and 3 h. Figure 5.15 compares the wear loss of the as-received MEZ and laser surface melted MEZ, as a function of time. It may be noted that wear loss is considerably reduced due to laser surface melting. The superior wear behaviour of laser surface melted MEZ is attributed to improved microhardness in the surface layer because of considerable grain refinement.

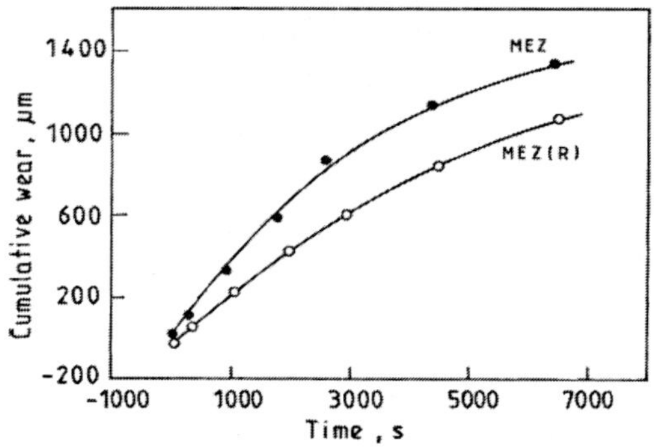

Figure 5.15: Comparison of the Wear Loss of MEZ & Surface Melted MEZ [12]

References

[1] Abundance and form of the most abundant elements in Earth's continental crust.

[2] B.L. Mordike, T. Ebert, "Magnesium Properties — applications — potential" *Materials Science and Engineering,* A302, 2001, pg. 37–45.

[3] National Research Council, *"Properties of Magnesium and Magnesium Alloys. In Trends in Usage of Magnesium",* Washington, DC: The National Academies Press, 1975, pg. 37-42.

[4] Magnesium Applications. International Magnesium Retrieved from: https://www.intlmag.org/page/mg_applications_ima

[5] P.K. Mallick (Ed.) Materials, *"Design and Manufacturing for Lightweight Vehicles",* Woodhead Publishing. Overview, 2010, pg. 1-32.

[6] Zhang, B., Liao, H., & Coddet, C, "Effects of processing parameters on properties of selective laser melting Mg–9% Al powder mixture", *Materials & Design,* 2012, 34, pg. 753–758.

[7] Gu Dongdong, Shen Yifu, "Effects of dispersion technique and component ratio on densification and microstructure of multi-component Cu-based metal powder in direct laser sintering", *J Mater Process Tech* 2007, 182: 564–73.

[8] Song, B.; Dong, S.; Deng, S.; Liao, H.; Coddet, C, "Microstructure and tensile properties of iron parts fabricated by selective laser melting", *Opt. Laser Technol.,* 2014, 56, pg. 451–460.

[9] Liu, S., & Guo, H, "Influence of hot isostatic pressing (HIP) on mechanical properties of magnesium alloy produced by selective laser melting (SLM)", *Materials Letters,* 265, 2020, 127463.

[10] The North American Die Casting Association, "Magnesium die casting handbook", vol. 201. NADCA; 1998. Pg. 1–78.

[11] Brandes EA, Smithells, *"Metal reference book".* 6th ed. London: Butterworths; 1983.

[12] Johnson Richard, "Friction stir welding of magnesium alloys", *Philip Threadgill. TMS;* 2003. Pg. 147–52.

[13] Dutta Majumdar, J., Galun, R., Mordike, B. ., & Manna, I, "Effect of laser surface melting on corrosion and wear resistance of a commercial magnesium alloy", *Materials Science and Engineering: A,* 361(1-2), 2003, pg. 119–129.

[14] Cui, Z., Shi, H., Wang, W., & Xu, B, "Laser surface melting AZ31B magnesium alloy with liquid nitrogen-assisted cooling", *Transactions of Nonferrous Metals Society of China,* 25(5), 2015, pg. 1446–1453.

[15] Labusch R., "A Statistical Theory of Solid Solution Hardening", *phys. Status solidi B* 41 (1970) 6-669.

[16] B. Beausir, S. Suwas, L.S. Tóth, K.W. Neale, J, "Fundenberger, Analysis of texture evolution in magnesium during equal channel angular extrusion", Acta Mater. 56, 2008, pg. 200–214.

[17] ASTM standard, *"Examination and Evaluation of Pitting Corrosion",* 1961, p. G46.

[18] J.E. Hilliard, in: R.T. Dehoff, F.N. Rhines (Eds.), *"Quantitative Microscopy",* McGraw-Hill, London, 1968, p. 45.

[19] ASTM standard, *"Standard Practice for Conducting Cyclic Potentiodynamic Polarization Measurements for Localized Corrosion",* 1961, p. G61.

[20] M.G. Fontana (Ed.), *"Corrosion Engineering",* McGraw-Hill, New York, 1987, p. 71.

[21] Cui, Z., Shi, H., Wang, W., & Xu, B, "Laser surface melting AZ31B magnesium alloy with liquid nitrogen-assisted cooling", *Transactions of Nonferrous Metals Society of China, 25*(5), 2015, pg. 1446–1453.

[22] Pu Z, Song GL, Yang S, Outeiro JC, Dillon O W Jr, Puleo DA, Jawahir IS, "Grain refined and basal textured surface produced by burnishing for improved corrosion performance of AZ31B Mg alloy", *Corrosion Science,* 2012, 57: pg. 192–201.

Chapter VI
Selective Laser Melting (SLM) of Titanium (Ti) Alloys

CHAPTER VI

6. Selective Laser Melting (SLM) of Titanium (Ti) Alloys

6.1. Introduction

Titanium is the ninth-most abundant element in Earth's crust (0.63% by mass) [1] and the seventh-most abundant metal. Titanium and its alloys are smart materials due to its exceptional high strength-weight ratio that is maintained at high temperatures and their outstanding resistant to corrosion. The most important use of titanium has been in the aerospace industry. However, the focus shift of market trends from military to commercial and aerospace to industry has also been reported. On the Other hand, titanium and its alloys are notorious for their poor thermal properties and are classified as difficult-to-machine materials [2]. These properties limit the use of these materials especially in the commercial markets where cost is much more of a factor than in aerospace. Machining is an important manufacturing process because it is almost always involved if precision is required and is the most cost effective process for small volume production. The two most useful properties of the metal are corrosion resistance and strength-to-density ratio, the highest of any metallic element [3, 4]. Like aluminium and magnesium, titanium metal and its alloys oxidize immediately upon exposure to air. Titanium readily reacts with oxygen at 1,200 °C (2,190 °F) in air, and at 610 °C (1,130 °F) in pure oxygen, forming titanium dioxide. Atmospheric passivation gives titanium excellent resistance to corrosion, almost equivalent to platinum. Titanium is capable of withstanding attack by dilute sulphuric and hydrochloric acids, chloride solutions, and most organic acids [5].

Titanium alloys are alloys that contain a mixture of titanium and other chemical elements. Such alloys have very high tensile strength and toughness (even at extreme temperatures). They are light in weight, have extraordinary corrosion resistance and the ability to withstand extreme temperatures. However, the high cost of both raw materials and processing limit their use to military applications, aircraft, spacecraft, bicycles, medical devices, jewellery, highly stressed components such as connecting rods on expensive sports cars and some premium sports equipment and consumer electronics.

Although "commercially pure" titanium has acceptable mechanical properties and has been used for orthopaedic and dental implants, for most applications titanium is alloyed with small amounts of aluminium and vanadium, typically 6% and 4% respectively, by weight. This mixture has a solid solubility which varies dramatically with temperature, allowing it to undergo precipitation strengthening. This heat treatment process is carried out after the alloy has been worked into its final shape but before it is put to use, allowing much easier fabrication of a high-strength product.

6.2. Properties of Titanium

Titanium's properties, which are a combination of high strength, stiffness, toughness, low density, and good corrosion resistance provided by various titanium alloys at very low to elevated temperatures, allow weight savings in aerospace structures and other high-performance applications.

6.2.1. Strength to Weight Ratio

The density is 4.506 g/cm^3 and the pure titanium is characterised for its specific strength, reaching up to strength of 590 MPa. When alloyed with other metals such as aluminium or vanadium, titanium becomes dramatically stronger than many steels. In terms of sheer strength, the best titanium alloys beat low to medium grade stainless steels. However, the highest grade of stainless steel is stronger than titanium alloys. We recommend sticking with a common titanium alloy if you're looking for strength. In alloy form, this strength extends dramatically up to 1250 MPa (exhibited in Grade Ti-15Mo-5Zr-3AI alloy). Its fatigue strength is about half of its tensile strength, and does not decline when exposed to welding or when submerged in seawater.

Titanium is a suitable component for applications that require a unique mixture of strength and material lightness. Tensile strength: Minimum 200 MPa to beyond 1300 MPa (varies depending on purity and alloy composition) [5].

Due to its impressive strength-to-weight ratio, titanium alloys are often used in strong products that need to be light, such as tennis rackets and bicycles. However, it is also used in ship hulls and propeller shafts due to its resistance to seawater. In terms of metal plating, electroplating services can be applied to titanium, for example platinum can be added to improve its appearance [6].

6.2.2. Corrosion Resistant

Because pure titanium readily reacts with oxygen, it naturally produces an oxide film that gives itself protection against corrosive materials and environments. It's corrosion resistant against chlorine compounds, seawater, common acids, and extreme temperatures. The use of titanium/palladium alloys virtually eliminates the risk of crevice corrosion in sea water. Titanium differs from most materials in that, if coupled to a more noble metal in an aggressive solution, the electrode potential of the titanium tends to be raised and the corrosion rate is reduced rather than increased [7]. Although titanium and its alloys are resistant to corrosion in many media, including aqueous solutions of chlorides, stress corrosion of commercially pure titanium and of titanium alloys can take place in a limited number of highly specific environments.

6.2.3. Refractoriness

While titanium is not classified as one of the five main refractory metals, it does have many of the common refractory metal characteristics, such as a high melting point (3035° F) and excellent resistance against corrosion. Titanium is also characterised by its refractory metal properties. Its melting point goes beyond 1650°C, significantly higher than aluminium and steel. Meanwhile, its coefficient of thermal expansion is 8.6 μm/ (m·K), lower than steel and copper [9]. As with most refractory metals, plating with titanium is not achievable with a conventional aqueous bath. It also cannot be plated by itself. One effective titanium plating, process involves the use of an alloy of titanium and nitrogen to form titanium nitride (TiN), which is deposited via either physical or chemical vapour deposition as opposed to traditional electroplating. Although a TiN coating is extremely thin, it is very hard and highly resistant to wear. The thinness also makes it easy to maintain desired tolerances during plating.

6.2.4. Biocompatibility

Titanium is considered the most biocompatible metal due to its resistance to corrosion from bodily fluids, bio-inertness, capacity for Osseo integration, and high fatigue limit. Titanium's ability to withstand the harsh bodily environment is a result of the protective oxide film that forms naturally in the presence of oxygen. The oxide film is strongly adhered, insoluble, and chemically impermeable, preventing reactions between the metal and the surrounding environment. Titanium was first introduced into surgeries in the 1950s after having been used in dentistry for a decade prior. It is now the metal of choice for prosthetics, internal fixation, inner body devices, and instrumentation. Titanium is used from head to toe in biomedical implants. One can find titanium in neurosurgery, bone conduction hearing aids, false eye implants, spinal fusion cages, pacemakers, toe implants, and shoulder/elbow/hip/knee replacements along with many more. The main reason why titanium is often used in the body is due to titanium's biocompatibility and, with surface modifications, bioactive surface. The surface characteristics that affect biocompatibility are surface texture, steric hindrance, binding sites, and hydrophobicity (wetting). These characteristics are optimized to create an ideal cellular response. Some medical implants, as well as parts of surgical instruments are coated with titanium nitride (TiN).

There is no known biological role for titanium. There is a detectable amount of titanium in the human body and it has been estimated that we take in about 0.8 mg/day, but most passes through us without being adsorbed. It is not a poison metal and the human body can tolerate titanium in large dose.

6.2.5. Health Effects of Titanium

There is no known biological role for titanium. There is a detectable amount of titanium in the human body and it has been estimated that we take in about 0.8 mg/day, but most passes through us without being adsorbed. It is not a poison metal and the human body can tolerate titanium in large dose.

Elemental titanium and titanium dioxide is of a low order of toxicity. Laboratory animals (rats) exposed to titanium dioxide via inhalation have developed small-localized areas of dark-colored dust deposits in the lungs. Excessive exposure in humans may result in slight changes in the lungs.

Effects of overexposure to titanium powder: Dust inhalation may cause tightness and pain in chest, coughing, and difficulty in breathing. Contact with skin or eyes may cause irritation. Routes of entry: Inhalation, skin contact, eye contact.

6.2.6. Environmental effects of Titanium

Low toxicity, When it is in a metallic powdered form, titanium metal poses a significant fire hazard and, when heated in air, an explosion hazard. No environmental effects have been reported [9].

6.3. Applications

Titanium in its pure form is a silvery metal known for its strength and low density compared to other similarly hard metals. In most industries, however, titanium alloy is much more commonly used. Because of its physical and chemical properties, the metal has become useful in a wide variety of industries and applications such as medical equipment, chemical plants, military installations, and sports gear.

6.3.1. Aerospace Industry

Due to titanium alloy's temperature and corrosion resistance, the material is used in the manufacturing of aircraft parts (airframe and fastening components), hydraulic system components, and landing gear. The Lockheed SR-71 Blackbird military aircraft is made primarily of titanium, and was touted for its speed and design efficiency. Although the stealth spy plane was retired in the late '90s, aircraft manufacturers continue to incorporate titanium in modern aerial vehicles to increase strength while reducing weight.

6.3.2. Marine Equipments

Titanium has a high level of corrosion resistance against seawater, making it a suitable component for ship rigs, propeller blades and shafts, and other parts submerged in water.

6.3.3. Sports Applications

The high strength-to-weight ratio of titanium lends itself to a range of applications in the sports industry. The material is used as components of sporting goods such as tennis rackets, baseball bats, golf clubs, bike frames, and ski equipment.

6.3.4. Medical Industry

Due to its inertness and non-toxicity, titanium is used in a wide array of medical applications including surgical implants, dental implants, surgical tools, and accessibility equipment.

6.3.5. Pyrotechnics

Powdered Titanium produces bright white sparks that are used in fireworks.

6.3.6. Pigment

Titanium dioxide – the natural oxide form of the metal – has found its use as a whitening agent in paints, plastics, and toothpaste.

6.3.7. Jewellery

The biocompatibility, high strength-to-weight ratio, and corrosion-resistant properties of titanium are useful in the jewellery industry for bracelets, rings, and necklace chains.

6.4. Densification Behaviour based on Laser Related Factors

Gu, D., Hagedorn et al [10] investigated the densification behaviour of commercially pure Titanium (CP Ti) powder (ASTM Grade 2, supplied by TLS Technik GmbH, Germany) was used in this study. The Ti powder had a spherical shape and particle size d_{50} = 19.5 μm and d_{100} = 43.3 μm using Fraunhofer ILT system which consist of mainly of a YLR-200 ytterbium fiber laser with a power of 200 W and a spot size of 70 μm (IPG Laser GmbH, Germany), an automatic powder layering apparatus, an inert argon gas with outlet pressure of 30 mbar as a protection system, and a computer system for process control.

Based on a series of preliminary experiments, the laser power (P) was optimized at 90 W. Meanwhile, the scan speeds (v) were set at 100, 200, 300 and 400 mm/s by the SLM control program, in order to change the processing conditions during one batch of experiment. Four different "linear energy densities" (LEDs) of 900, 450, 300, and 225 J/m, which were defined by [11]:

$$\text{Laser Energy Density (LED)} = \frac{\text{Laser power}}{\text{scanning speed}}\ \text{J/m} \qquad (6.1)$$

The cross-section microstructures of Ti parts processed using various SLM parameters are provided in figure 6.1. Layer wise microstructural features were generally observed, due to the layer-by-layer incremental deposition manner of the SLM process.

At a relatively low v of 100 mm/s and a high LED of 900 J/m, thin, narrow micro cracks with an average length of 400 μm were present between the neighbouring layers (figure 6.1a), producing 3.3% porosity in the SLM part. Dense, metallurgically bonded layers without any interlayer pores or cracks were formed as the scanning speed increased to 200 mm/s. The as-solidified molten pools generally had a clear configuration, typically showing a stable solidification front (figure 6.1b). At a higher scanning speed of 300 mm/s, although the cross-section microstructure of SLM part was still fully dense, the distribution state of the layers became uneven. Optical microscopy images showing interlayer microstructures on cross-sections of SLM-processed Ti parts at different processing parameters: as shown in figure 6.1(a), (b).

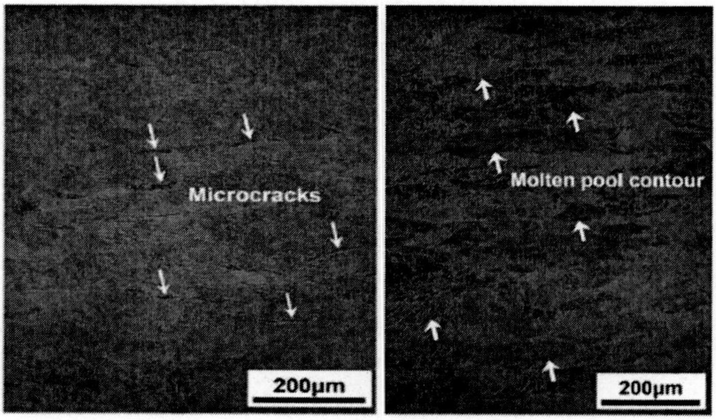

Figure 6.1: (a) 900 J/ m, 100 mm/s Figure 6.1: (b) 450 J/ m, 200 mm/s;

Microstructural Image with Laser Energy Density and Scan Speed [10]

The structures of solidified molten pools were elongated along the laser scan direction (figure 6.2 a), implying a significantly elevated thermodynamics of the pool as the scan speed increased.

In these two instances, the obtainable densification rate of SLM parts generally approached 99.5%. At an even higher v of 400 mm/s and attendant low LED of 225 J/m, irregularly shaped micro pores with a mean size of ~30 μm were formed in the interlayer areas (figure 6.2b), thereby increasing the residual porosity of the SLM part to 5.7%.

The micro cracks formed at a low v and attendant high LED (figure 6.1a) are believed to be thermal cracks caused by residual thermal stresses. During the SLM densification process, shrinkage occurs rapidly once the powder material becomes fully molten. A nominally flat powder bed surface becomes a curved surface or a recessed region due to the effect of densification caused by conversion of loose powder to dense liquid.

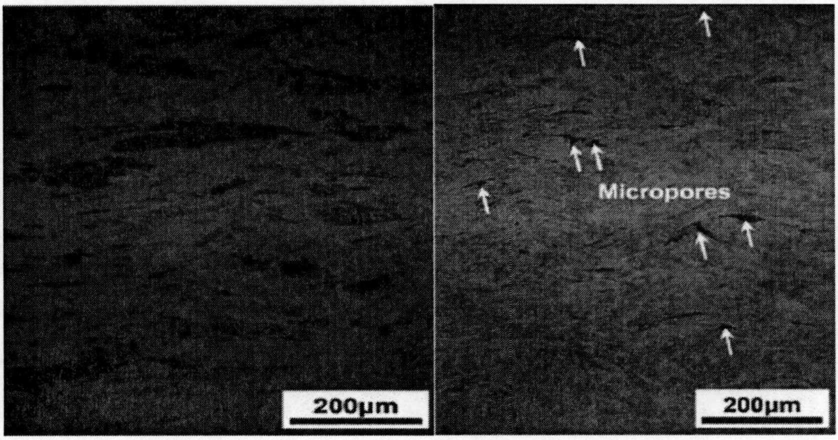

Figure 6.2: (a) 300 J/ m, 300 mm/s Figure 6.2: (b) 225 J/ m, 400 mm/s

Microstructural Image with Laser Energy Density and Scan Speed [10]

Uhlmann, E et al [12] investigated the effect of Ti alloy in aircraft components using SLM process and the density of the specimens is discussed with other mechanical properties. He used TiAl6V4 with a chemical composition of Ti(91.23 %wt), Al(5.21% wt), V(3.47% wt) and Fe(0.10%wt) determined by EDX analyses[13] which delivers the same which is suitable for SLM process.

The particle shape has been also determined by the Scanning Electron Microscopy (SEM) LEO 1455 VP according to DIN EN ISO 3252 [14]. The particles have different shapes depending on the manufacturing process. In this case, the powder results from inert gas atomization. In this spherical shaped which gives good flowablility and high packing density of powder as shown in figure 6.3. The flowability also depends on the grain size distribution. A Sieve Analysis has been carried out according to DIN 66165-1 with the Test Sieve Shaker Haver EML 200 digital plus [15]. The IPK analysis shows a Gaussian distribution which is more suitable than the given manufacturer's data. The Grain Size Distribution is applicable for SLM.

The residual moisture was measured with the Moisture Analyzer Radwag MAC 110. The maximum acceptable value is 0.025 %. The average detected residual moisture was around 0.017 %. Therefore the residual moisture is suitable for SLM.

The Packed Filling Density (PFD) is detected because the machine laser parameters refer to a certain packing density of the metal powder and has an influence on the weldability. This has been analysed according to DIN EN ISO 3923 [16]. The PFD is calculated by dividing mass by volume of the powder material. In this case the Scott voltmeter procedure has been used.

Figure 6.3: TiAl6V4 Powder Particles, 110x (left) and 1300x (right) [12]

A high packing density leads to less blowholes and bubbles hence a better microstructure. The determined filling density is around 2.35 g/cm^3 which equates to 53 % of the raw material density of 4.43 g/cm^3. This PFD is suitable for SLM. From experiences by working with other materials at the Fraunhofer IPK, a PFD of 50 % to 60 % has been proven to be acceptable.

The production of TiAl6V4 specimens was carried out using a SLM 250HL machine by the company SLM Solutions GmbH, Lübeck, Germany. The significant parameters used for the manufacturing of testing geometries are the laser power P_L, the scanning velocity v_s and the focus diameter d_s, which depend on the focus position. The used SLM parameters for the production of testing geometries are shown in Table 6.1. The layer thickness of 50 µm remains constant during the whole manufacturing process. The hatch parameters used for the volume area as shown in table 6.2 with a strategy of chess board.

Table 6.1: Process Parameters for SLM Process [12]

Process parameters	Volume hatch	Volume contour	Supports
Focus position X_f (mm)	2	1	0
Laser power P_L(W)	275	100	175
Scanning velocity V_s(mm/s)	975	400	470

Table 6.2: Hatch Parameters [12]

Hatch Parameters	Volume
Panel width [mm]	5
Panel height [mm]	5
Panel overlap [mm]	0
Min. panel size [mm]	2.5
Contour spacing [mm]	0.15
Incrementing angle [°]	67

Round specimens in 0°, 45° and 90° orientation are generated according to VDI 3405-2. Figure 6.4 shows the exposure in the SLM-process and the generated specimens.

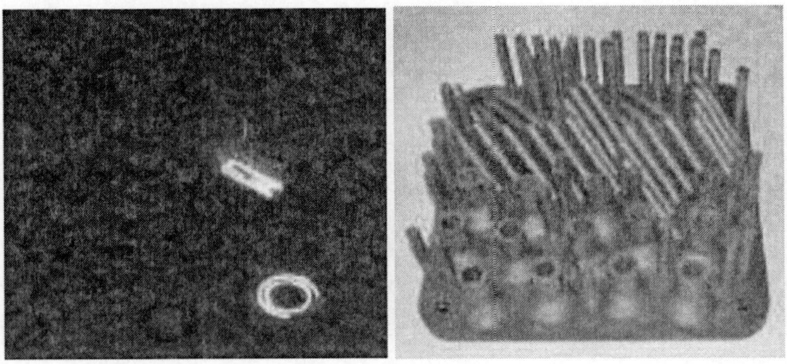

Figure 6.4: SLM-exposure of the Specimens (left) and Generated Specimens on the Substrate Plate (Right) [12]

A thermal post-processing is used before the hot isostatic pressing (HIP) in order to increase the surface quality, close micro cracks and reduce very high residual stresses in the component. According to DIN 17869 stress relief heat treatment is recommended for internal welding stresses of multi-layered weld seams in a component. The process parameters are based on indications of DIN 65083 for thermal treatment of casted components made of titanium and titanium alloys for aerospace [17, 18]. The following process parameters were used as:

- Stress relief heat treatment under protective gas atmosphere or in vacuum.
- Holding period: 60 min.
- Temperature: 675 °C.
- Cooling: with open stove under inert gas.
- Cooling rate: ≤2.5 K/min.

Hot Isostatic Pressing according to DIN 65083 "causes the healing up of internal structural defects such as micro blowholes and pores in castings, through annealing at high temperatures and pressures". So the HIP-process reduces the porosity and increases the density of the part. Thus fatigue properties get improved. The static and dynamic strength, the breaking elongation and durability are increased and more uniform mechanical properties are achieved [19, 20]. Table 6.3 shows the process parameters according to DIN 65083.

Table 6.3: Process Parameter HIP for TiAL6V4 [15]

HIP parameters	**DIN65083 Used parameters**	**Used parameters**
Pressure(MPa)	-	100
Tempereature(°C)	900-920	910
Holding time(min)	120	135
Cooling rate((°C/hr)	≤150	120
Heating rate((°C/hr)	-	300
Gas	Inert gas	Argon
Klin furniture	-	Graphite crucible with Al_2O_3 base
Cooling	-	Furnace cooling, up to 400°C under inert gas

Archimedes' principle was applied to find the density of parts. The weight shown on the balance is equal to the weight of the displaced fluid, caused by the solid body [21, 22]. As reference, the literature density of the casted material without porosity, the value of 4.43 g/cm^3 is used [17]. The detected density of the SLM-specimen is around 4.35 g/cm^3 which correspond to 98.19 % relative density.

6.5. Mechanical Properties of SLMed Part

6.5.1. Hardness

Gu, D., Hagedorn et al [10] investigated the hardness along with the densification behaviour with respect to various process related factors as discussed in section 6.4. The wear performance can be obtained by using the various scan rate and laser power to get mere dense part. Figure 6.5 depicts the nano indentation load–depth curves measured on the polished sections of SLM-processed Ti parts. The indentation depths of the parts processed at a low v of 100 mm/s (1.2762 μm) and a high v of 400 mm/s (1.1389 μm) were larger than those of the parts prepared using the optimal scan speeds (1.0168 μm at 200 mm/s and 0.9869 μm at 300 mm/s). The corresponding dynamic nano hardness Hd values (2.32 GPa at 100 mm/s and 2.92 GPa at 400 mm/s) were considerably lower than those obtained at scan speeds of 200 mm/s (3.66 GPa) and 300 mm/s (3.89 GPa). Nevertheless, the SLM-processed Ti parts, for all the given processing parameters, demonstrated superior hardness property compared to conventional powder metallurgy (PM) processed Ti with a typical hardness of ~1 GPa [26]. In general, residual stresses are quite large in layer-by-layer fabricated SLM parts, but residual stresses are not always disadvantageous [23].

On the premise of a sufficiently high densification without the formation of cracks or pores (figure 6.1a and 6.2 a), the retention of a reasonable level of residual stress in SLM-processed parts favours the enhancement of hardness. Moreover, the significant grain refinement effect due to laser rapid solidification favours a further increase in the obtainable hardness.

Uhlmann, E et al [12] investigated the effect of Ti alloy in aircraft components using SLM process and the density of the specimens is discussed in section 6.4. Uhlmann, E investigated the mechanical properties of Ti Al4V6 alloy i.e. hardness. The Vickers micro hardness is used for the hardness analysis as the process can be used for almost every material and small specimens. The measurable Vickers Hardness range is 10 HV to 2000 HV. DIN 6507-1 describes the implementation of the process [24]. The measurement is made in the xy- and xz-direction. The determined hardness in xy-direction is 316 HV30 and in xz-direction 320 HV30. Hence the hardness of the specimens are approximately 88 % and 89 % in reference to the average literature value of 360 HV30, as the literature hardness lies between 330 to 390 HV30 [25].

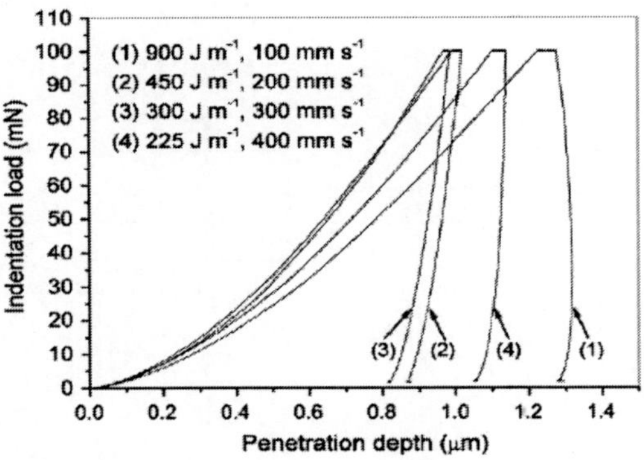

Figure 6.5: Loading–unloading Curves [10]

6.5.2. Fatigue

Mario Viespoli, L., Bressan et al [27] with his team investigated the Creep and high temperature fatigue performance of Ti-based 6Al-4V titanium alloy using SLM process.

The specimens used for the present study were manufactured by selective laser melting using a SLM Solutions (SLM 280) device. The build direction coincides with the longitudinal axis and the process details are indicated in table 6.4.

Table 6.4: Parameters for SLM Processed Ti6Al4V (Ti Alloy Part) [27]

Device	SLM 280 HL – SLM solutions
Laser power[W]	275
Laser scan speed[mm/s]	1100
Hatch style	Strip hatch
Hatch rotation[°]	15
Hatch distance[μm]	120
Hatch length[mm]	10
Atmosphere	Argon
Materia	Ti6AL4V

The geometries produced, for fatigue is reported in figure 6.6. During SLM fabrication a defect was introduced in the form of a slight misalignment of the build axis in correspondence of the minimum cross section.

The fatigue specimens were produced to a minimum diameter of 6 mm. Four of them were being tested at room temperature with the as built surface, while the rest of them had the surface lathe-machined down to a diameter of 5 mm as reported in figure 6.6.

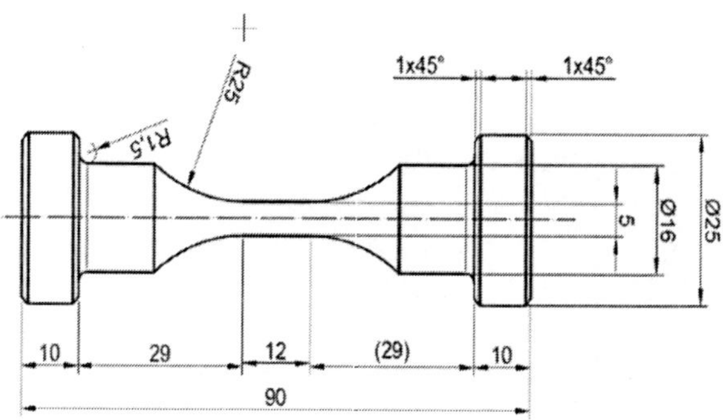

Figure 6.6: Fatigue Specimen Processed by SLM [27]

The creep specimens where tested with the as-built surface. All the specimens were tested with the as-built microstructure consisting of a circular α' martensitic phase in elongated β grains, certain degree of porosity can be detected in the longitudinal cut as observed from figure 6.7.

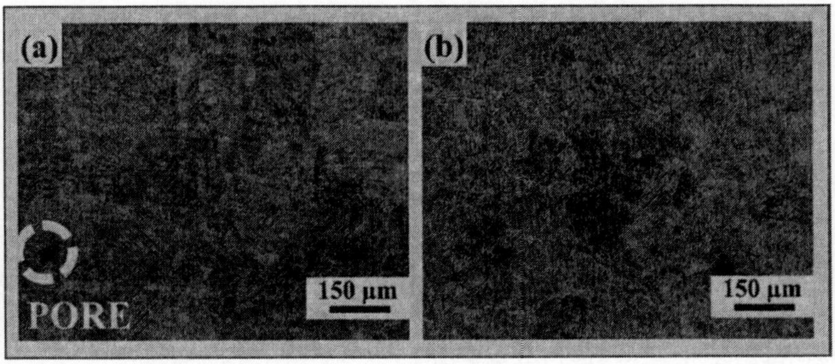

Figure 6.7: (a) & (b) Ti6Al4V SLM Microstructure along Build Direction [27]

The fatigue testing was performed, by a hydraulic machine, in strain control applying a fully reversed displacement that corresponds to a displacement ratio - 1, at a strain rate of 0.1%/s with the feedback provided by an induction axial extensometer and their results are reported in figure 6.8.

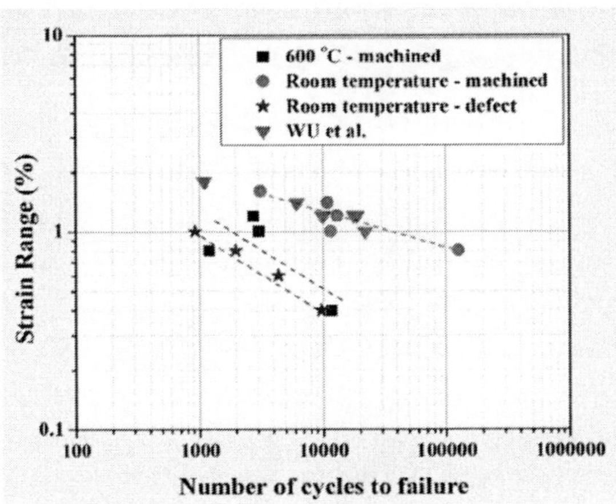

Figure 6.8: Fatigue Results [28]

The extensometer, of the kind suited for high temperature testing, had ceramic pins clipped in small pits practiced at a distance of 10 mm in the central section of the specimen. Three series of specimens were fatigue tested: four at room temperature with the as-built surface, six at room temperature with machined surface and four at high temperature and machined surface after having been warmed up to 600 °C for 30 min by an induction coil. The temperature was detected by a thermocouple welded to the fillet radius area, in order to not weaken the neat cross section by practicing a weld. The temperature in the centre area was correctly known by having calibrated the correlation between the temperatures in the two points on a dummy specimen.

The tests performed at room temperature on machined surface specimens show a behaviour which is in perfect agreement with analogous tests in the literature which was disused in reference[28].

Assessed so the reproducibility of the results, the other two series show the strong weakening produced by a surface defect and by an increase of temperature. The specimens tested with an as-built surface in presence of the defect have all experienced failure with cracks starting from the defect, see figure 6.9a, and a fatigue life noticeably inferior to the machined specimens.

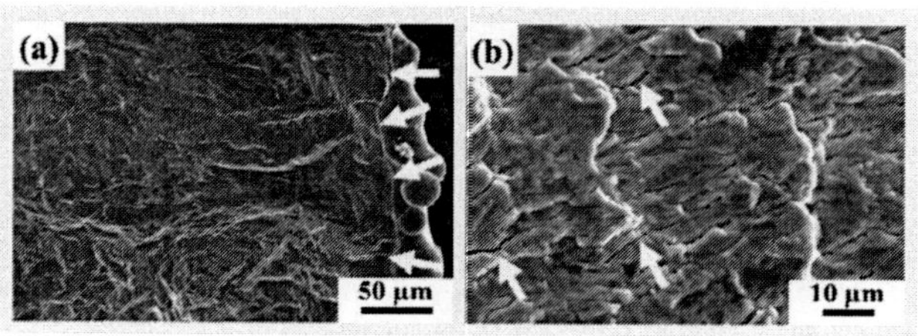

Figure 6.9: (a) Fatigue Crack Initiation at a Surface Defect and Partially Melted Particles, (b). Secondary Cracking in High Temperature Fatigue Fracture [27]

Considered the entity of the defect, a misalignment of the build axis of 75 μm, it is evident how even a small defect can have huge repercussions on the structural integrity, which is particularly of importance for the production of complicated topology-optimized geometries, the realization of which is allowed by the great flexibility of the AM technologies, where postproduction surface machining can be non-cost effective or impossible. A similar reduction of strain-controlled fatigue life is originated by an increase of ambient temperature to 600 °C. The fatigue failure at high temperature is characterized by a diffused secondary cracking originated by weakening of the grain boundary regions, as seen in figure 6.9b.

Rafi, H. K et al [29] and his team members investigated the microstructural and Mechanical Properties of Ti6Al4V Parts Fabricated by Selective Laser Melting (EOS M270 SLM machine) and Electron Beam Melting (Arcam S400 EBM machine). The process parameters used is with respect to [30, 31] were used for the process. The average particle size is 36μm for EOS and for Arcam is 60μm.Cylindrical specimens are made as per ASTM: E8 standards for tensile testing and fatigue testing specimens were fabricated as shown in figure 6.10(a) & (b).

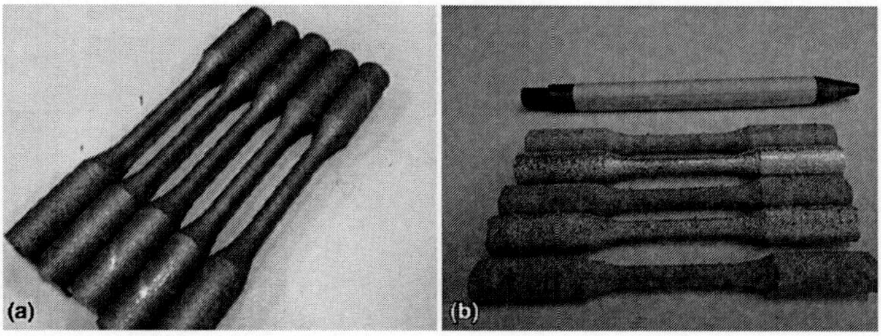

Figure 6.10: (a) & (b) Specimen Produced in SLM and EBM Respectively [29]

S-N curves illustrating the fatigue behaviour of SLM produced Ti64 and EBM-produced Ti64 are shown in figure 6.11. For SLM-produced Ti64 samples a fatigue limit of 550 MPa was observed (figure 6.11a). A comparison with MMPDS (Metallic Materials Properties Development and Standardization Handbook, Battelle Memorial Institute, Columbus, Ohio, April 2010) fatigue data shows better fatigue performance for Ti64 specimens built using SLM compared to Ti64 that is cast and annealed (430 MPa).

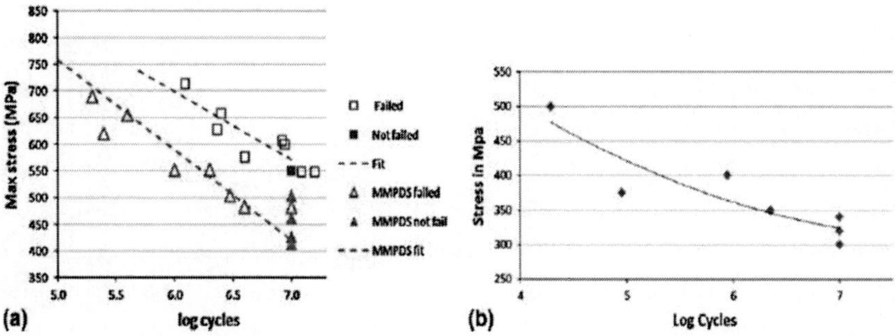

Fig. 6.11: S-N Curve of Ti64 Samples (a) SLM and (b) EBM [29]

However, the fatigue performance of EBM-produced Ti64 samples was inferior (figure 6.11b, fatigue limit: 340 MPa) as compared to the SLM produced Ti64 and the MMPDS data. The better fatigue strength properties of SLM-produced Ti64 can be attributed to the martensitic phase. Martensite impedes dislocation motion and thus leads to a strengthening effect. Because of this strengthening effect, the total strain amplitude introduced during fatigue testing causes smaller plastic strain.

Leuders, S et al [32] investigated made several Ti-6-4 specimens with standardized geometries were employed, i.e. cylindrical specimens for high-cycle fatigue (HCF) testing and compact tension (CT) specimens for investigation of crack growth. The specimens were processed using a SLM 250HL machine (SLM Solutions GmbH) in combination with SLM AutoFab software (Marcam Engineering GmbH) employing an yttrium fibre laser with a maximum power of 400 W. Layers of 30 µm thickness were fabricated using Ti-6-4 powder with an average particle size of 40 µm. The platform was heated to 100 °C. Chemical analyses by means of X-ray fluorescence analysis of the Ti-6-4 powder revealed oxygen and nitrogen contents of 0.14% and 0.04%, respectively. Following SLM processing both elements were not enriched significantly, and 0.17% of oxygen and 0.06% of nitrogen were determined. Four different microstructural conditions of the Ti-6-4 alloy were considered. The one without heat treatment is referred to as "As-built", the other samples were heat treated in vacuum or under Argon atmosphere at 800 °C (below b-transus) and 1050 °C (above b-transus) for 2 h or hot isostatically pressed (Table 6.5 in section 6.5.4). All specimens were furnace cooled. For an optimized temperature control three different furnaces were used. A Nabertherm N30/85HA furnace equipped with an inert gas chamber was used for the 800 °C heat treatment. For the 1050 °C heat treatment a Linn HK 40.27 furnace was used and the specimens were encapsulated in evacuated quartz glass tubes. The HIP treatment was performed by Bodycote European Holdings GmbH CEG Technology Group under argon gas 2 hours.

The SLM-specimens were subjected to uni-axial sinusoidal cyclic loading with a stress ratio of $R = -1$, a test frequency of $f = 40$ Hz and at a stress amplitude of 600 MPa. All experiments were carried out at room temperature. The results of the HCF fatigue tests clearly show that a post treatment had a strong impact on the fatigue behaviour. The mean fatigue life ranged from 27,000 to 290,000 cycles to failure for the as-built and the heat-treated samples. Inferior fatigue lives were observed for the as-built condition with a mean fatigue life of 27,000 cycles, the Ti-6-4 heat treated at 800 °C showed mean fatigue lives of 93,000 cycles, the 1050 °C heat treated condition 290,000 cycles. By contrast, none of the HIP SLMed specimens failed during these tests, which were interrupted at 2×10^6 cycles. Due to their improved performance in the HCF regime, the fatigue limit for HIP SLMed Ti-6-4 was determined using the staircase method employing a total of 18 samples. For data analysis the procedure proposed by Huck [33] was used, and a fatigue limit of 620 MPa with an estimated standard deviation of 5.4 MPa was determined. Analyzing data according to Dixon and Mood [34], the fatigue strength is calculated to be 630 MPa with an estimated standard deviation of 5.3 MPa.

6.5.3. Creep

Mario Viespoli, et al [27] with his team investigated the creep on Ti-based 6Al-4V titanium alloy using SLM process along with high temperature fatigue performance which was discussed in section 6.5.2.

The creep testing was performed on a series of 8 specimens of the geometry specified in figure 6.12 with the parameters discussed in table 6.4 in section 6.5.2. The samples were tested applying a constant nominal stress with a drop weight machine with induction heating at temperatures of 450, 550 and 650 °C, respectively.

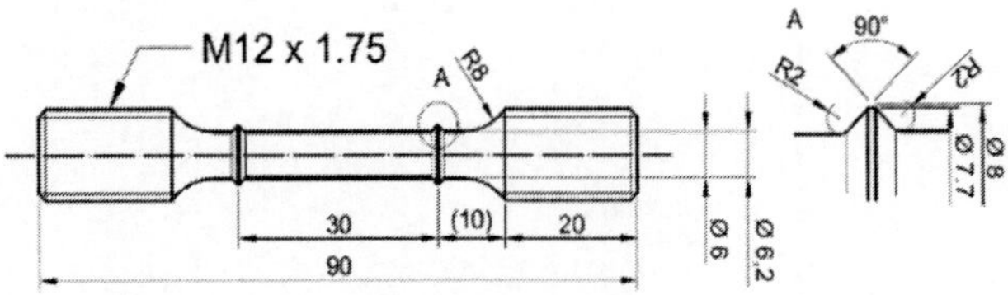

Figure 6.12: Creep Specimen Processed by SLM [27]

The results are reported in Fig. 6.13 in terms of nominal stress vs. time to failure, showing a consistent behaviour for the whole range of time to failure, from 120 s to 655 hours.

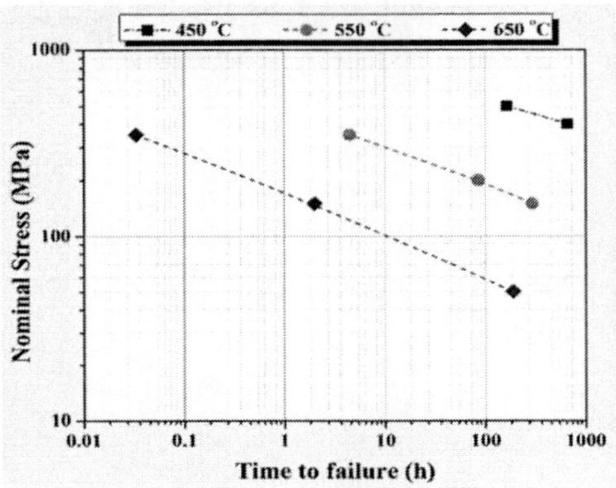

Figure 6.13: Creep Testing: Nominal Stress vs. Time to Failure [27]

The consistency of the behaviour is evident in the summary of the results in terms of nominal stress versus Larson Miller parameter, in figure 6.14, in which all the results can be approximately described by a single power function. Observing the steady state creep behaviour reported in figure 6.15 as applied stress versus strain rate, the stress exponent shows a reduction with increased temperature and lower applied stress, thus indicating an increased importance of the diffusive creep deformation versus the dislocation dominated creep, which is predominant for lower temperatures and higher stress levels.

Comparing the results obtained, the creep testing results available in the literature in hot forged and forged and annealed Ti6Al4V alloy, the as built SLM material has an analogous performance in terms of Larson Miller parameter vs. applied stress [35, 36].

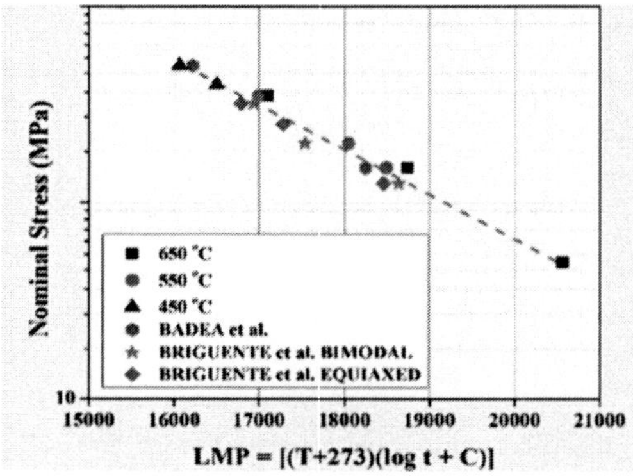

Figure 6.14: Creep Nominal Stresses vs. Larson Miller Parameter [35, 36]

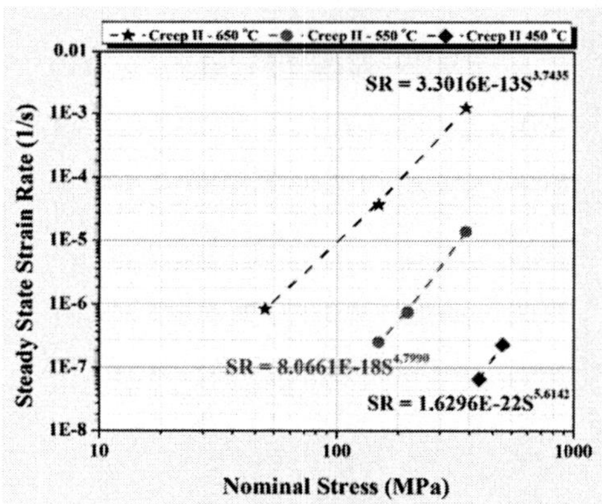

Figure 6.15: Secondary Creep Strain Rate vs. Nominal Stress [27].

Figure 6.16 reports the correlation between the applied stress and the ductility for the three test temperatures. The nominal strain was measured during the creep tests by the means of an extensometer attached to the two rings, which delimitate the uniform diameter section of the creep specimens. Despite the important scatter of the results, it is evident that the applied stress has a determinant influence on the elongation to failure, with values ranging from 75% at a nominal stress of 50 MPa at 650 °C down to 20% for 500 MPa applied at a temperature of 450 °C.

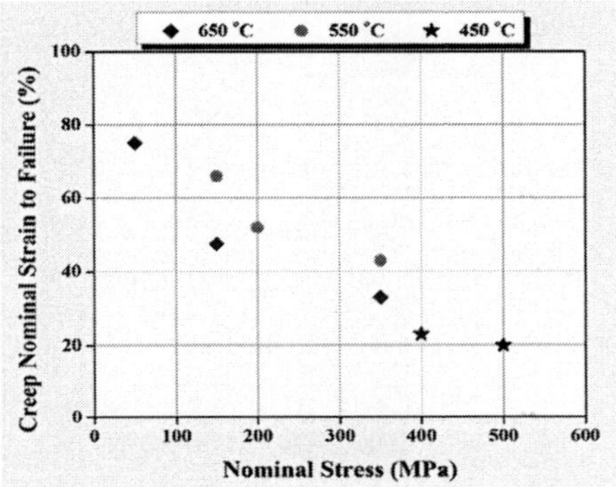

Figure 6.16: Creep Test: Influence of Stress on Ductility at different Temperatures [27]

Figure 6.17 presents the SEM magnification of the fracture surfaces of two creep specimens, tested at 650 °C, 50 MPa and 450 °C, 500 MPa.

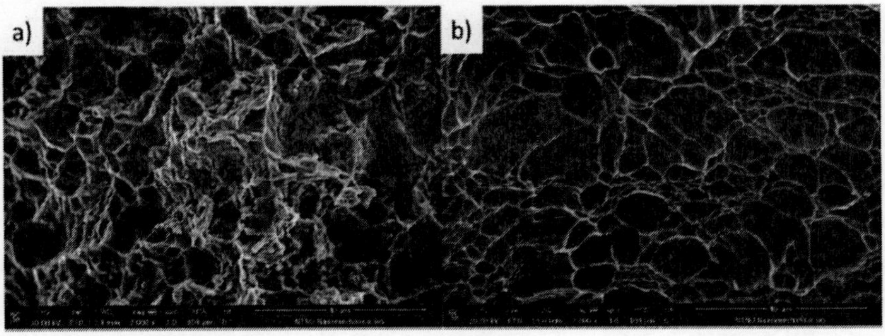

Figure 6.17: Fracture surface of Specimen Creep Tested at: (a) 650 °C, 50 MPa (b) and 450 °C, 500 MPa [27]

The failures are characterized by pore nucleation and growth, but they present a different aspect according to the test duration, with a much rougher surface in case of higher T and lower stress. This is possibly due to material oxidation.

6.5.4. Tensile Strength

Rafi, H.K et al [29] and his team members investigated the microstructural and Mechanical Properties of Ti6Al4V Parts Fabricated by Selective Laser Melting (EOS M270 SLM machine) and Electron Beam Melting (Arcam S400 EBM machine). The process parameters used is with respect to [37, 38] were used for the process. The average particle size is 36μm for EOS and for Arcam is 60μm. Cylindrical specimens are made as per ASTM: E8 standards for tensile testing and fatigue testing specimens were fabricated as shown in figure, which was shown in section 6.5.2.

Tensile results of EBM- and SLM-produced Ti64 samples are summarized in Table 6.5. Corresponding stress-strain curves are shown in figure 6.18 & 6.19. The results are the mean values based on five duplicate tests. The yield strength, ultimate tensile strength, and percentage strain were obtained as direct output from the tensile testing machine.

Table 6.5: Tensile Results for SLM and EBM Produced Ti Alloy Samples [29]

Description	**Stress at yield, (Offset 0.2%), MPa**	**Ultimate tensile stress, MPa**	**Strain at break, %**
EBM (vertically built and Machined)	869 (SD: 7.2)	928 (SD: 9.8)	9.9 (SD: 1.7)
SLM (vertically built and Machined)	1143 (SD: 30)	1219 (SD: 20)	4.89 (SD: 0.6)
% Increase	31	31	-50
EBM (horizontally built and Machined)	899 (SD: 4.7)	978 (SD: 3.2)	9.5 (SD: 1.2)
SLM (horizontally built and Machined)	1195 (SD: 19)	1269 (SD: 9)	5 (SD: 0.5)
% increase	33	30	-47
ASM Handbook [10] (cast and annealed)	885	930	-

The percentage strain-to-failure was measured using a clip-on extensometer that was attached to the gage section of the test specimen. Substantial difference in tensile properties can be seen between the EBM-produced Ti64 and the SLM-produced Ti64 samples. However, the tensile strength values are comparable with or even better than the standard Ti64 material data given in the ASM Handbook [39]. The tensile results are due to the differences observed in the microstructures. The higher tensile strength observed in SLM-produced Ti64 can be attributed to the martensitic microstructure as compared to the lamellar structure in EBM-produced Ti64 samples. Vilaro et al. [40] and Facchini et al. [41] also reported higher yield and ultimate strengths for as-fabricated SLM Ti64 specimens. Though the yield strength and tensile strength of EBM-produced Ti64 samples are low when compared to the SLM-produced Ti64 samples, the strain at break is higher indicating good ductility. The tensile test results are in conformation with the bulk hardness tests. The Rockwell hardness test resulted in higher hardness for SLM-processed Ti64 (HRC: 41) compared to the hardness of EBM-processed Ti64 (HRC: 33). Martensitic microstructure in Ti64 results in high strength and low ductility. Because of the low ductility, the SLM Ti64 samples failed at lower strain values. Ductility of SLM produced Ti64 samples can be improved by proper post-heat treatment which decomposes the harder martensitic phase to softer α phase [42].

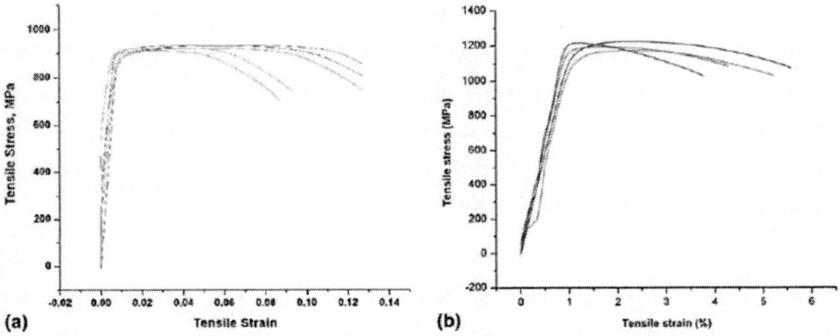

Figure 6.18: Stress-Strain Plots of Ti64 Samples (a) EBM Specimen, (b) SLM Specimen, Built in Vertical Orientation [29]

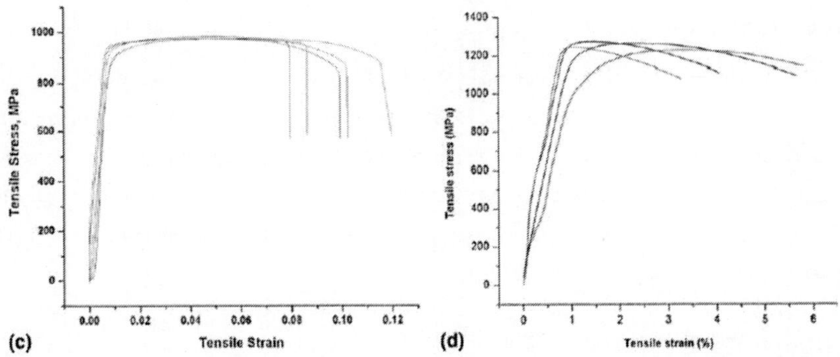

Figure 6.19: Stress-Strain Plots of Ti64 Samples (a) EBM Specimen, (b) SLM Specimen, Built in Vertical Orientation [29]

6.6. Tribological Behaviour of SLMed Parts

6.6.1. Wear

Gu, D. et al [10] investigated the wear performance along with the hardness and densification behaviour with respect to various process related factors as discussed in section 6.4. The wear performance can be obtained by using the various scan rate and laser power to get mere dense part.

Figure 6.20 depicts the variations of COF and wear rate with sliding time for all four SLM-processed Ti samples. Detailed SEM studies of the corresponding worn surfaces are shown in figure 6.21 and the underlying mechanisms for the formation of various topographical features were proposed to be as follows. In general, the coefficient of friction (COF) values for each specimen exhibited relatively steady-state behaviour with few local fluctuations. Nevertheless, the applied SLM conditions for various Ti samples exerted a significant influence upon the obtainable wear performance. At a relatively low v of 100 mm/s, the average COF reached a high value of 1.41 (figure 6.20 a), resulting in a considerably elevated wear rate of 1.19 × 10^{-3} mm^3 N^{-1} m^{-1} (figure & b). The worn surface primarily consisted of parallel, deep grooves representing abrasion wear. The presence of loose fragments at the edges of grooves revealed the local severe deformation and plowing of the surface during sliding.

Some irregular shaped debris in the form of particles was also observed on the wear surface (figure 6.21a). On increasing scanning speed to 200 mm/s, the worn surface showed much shallower grooves, free of any loose abrasive fragments (figure 6.21b), thereby reducing the mean COF and attendant wear rate to 1.16 and 8.82 × 10^{-4} mm^3 N^{-1} m^{-1}, respectively (figure 6.20). Interestingly, using a reasonable scanning speed of 300 mm/s, the worn surface of Ti sample became rather smooth and was found to be covered with adhesion tribo layer (figure 6.21c).

The lowest COF of 0.98 and wear rate of 8.43×10^{-4} mm^3 N^{-1} m^{-1} were accordingly obtained (figure 6.20). In this situation, due to the full densification response, the formation of refined a α' martensitic microstructure, and the enhancement of hardness (figure 6.19), the protective tribo layer can easily smear plastically on the surface being slid. It is thus reasonable to consider that the mechanism of material removal during sliding changes from abrasion (figure 6.21a) to adhesion of the tribo layer (figure 6.21c).

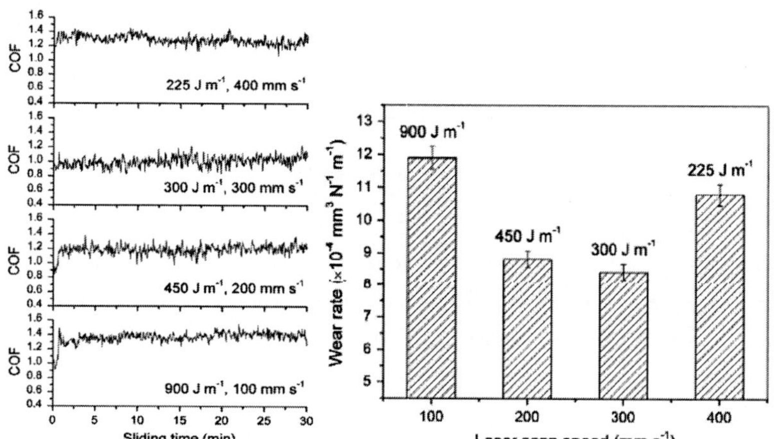

Figure 6.20: (a) COF and (b) Wear Rate of SLM-processed Ti Parts under different Processing Conditions [9]

Such a transition favours the reduction in wear rate after sliding. For SLM Ti sample processed at an even higher v of 400 mm/s, although the localized formation of tribo layer was still observed on the worn surface, severe spalling and delamination of the tribo layer occurred (figure 11d). The obtained COF value 1.29 and wear rate 1.08 $\times 10^{-3}$ mm^3 N^{-1} m^{-1} showed a significant increase in this case (figure 6.20).

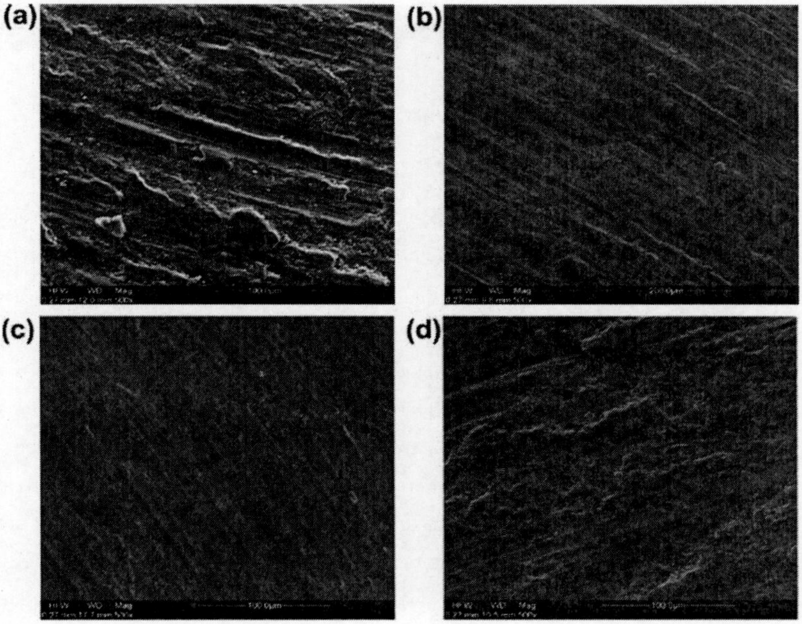

Figure 6.21: SEM Images of Worn Surfaces of SLM-processed Ti Parts with various Process Parameters [10]

6.7. Defects in SLM of Titanium Alloys

6.7.1. Crack

Leuders, S et al. [32] investigated made several Ti-6-4 specimens with standardized geometries were employed, i.e. cylindrical specimens for high-cycle fatigue (HCF) testing and compact tension (CT) specimens for investigation of crack growth. The specimens were processed using a SLM 250HL machine (SLM Solutions GmbH) in combination with SLM AutoFab software as discussed in section 6.5.2. The results obtained from crack growth measurements employing the potential drop technique are shown in figures 6.22, which clearly shows the crack growth curves for SLM-processed Ti-6-4 in different conditions.

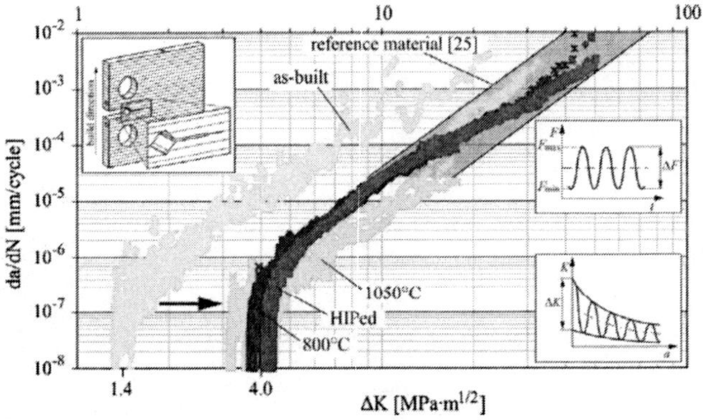

Figure 6.22: Crack Growth Curves for SLM-processed Ti-6-4 in different Conditions, Crack Growth Perpendicular to Build Direction [32]

Crack growth is perpendicular to the build direction. Data for conventionally processed reference material 25 as [43] are depicted by the scatter band shown in the upper right. Insets depict the CT specimen including information on SLM building direction and schematics of the testing procedures used in the different crack growth regimes. The threshold value for the conventionally processed reference material from literature is 4 MPa $m^{1/2}$[20]. As the two sets of curves show, there is a fundamental difference between the as-built and the heat treated (including HIPed) material. In figure 6.22 the direction of crack growth is perpendicular to the build direction, i.e. the orientation of the crack plane is parallel to the layer plane during SLM. Looking at the results for the as-built components it is obvious that there is a high variation in crack growth rate for a given value of cyclic stress intensity factor DK in the whole range tested.

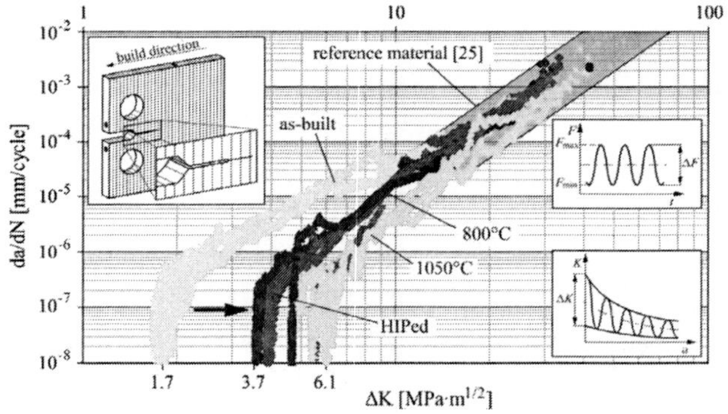

Figure 6.23: Crack Growth Curves for SLM-processed Ti-6-4 in different Conditions, Crack Growth Parallel to Build Direction [32]

The average threshold value for as-built material is 1.4 MPa $m^{1/2}$. The results for specimens heat treated at 800 °C as well as for the HIP SLMed specimens are very similar, and the scatter is minimized. By contrast, the data for specimens treated at 1050 °C again show larger scatter in the threshold area as well as in the Paris regime. The mean threshold value for the three considered treatments is equal to 3.9 MPa $m^{1/2}$ with a standard deviation of ±0.4 MPa $m^{1/2}$ (10%). These results are very close to the values reported in literature for the reference material. Figure 6.23 shows the results for crack growth parallel to the build direction. Looking at the curves of as-built samples, it can be seen, that the crack growth behaviour at lower crack growth rates is similar to the data presented in figure 6.22 for as-built material. Only scatter seems to be decreased here. The average threshold value is equal to 1.7 MPa $m^{1/2}$. At higher crack growth rates, however, the curves obtained for as-built specimens coincide with both the data obtained for heat treated material and the reference material. After heat treatment a threshold value between 3.7 MPa $m^{1/2}$and 6.1 MPa $m^{1/2}$ has been determined. The mean threshold value and the standard deviation for these three treatments can be described by a value of 4.6 ± 0.9 MPa $m^{1/2}$ (20%). Clearly, a significant improvement in crack growth behaviour can be achieved by an appropriate treatment. This effect is evident in the shift of the curves to the right in figures. 6.22 and 6.23 to higher levels of the cyclic stress intensity factor ΔK.

6.7.2. Porosity

Uhlmann, E et al [12] investigated the effect of Ti alloy in aircraft components using SLM process as discussed in section 6.5.1 along with the analysis of HIP SLMed and non-HIP SLMed specimens using Computed tomography (CT). Metronom 800 by Carl Zeiss was used for the analysis.

Here cylindrical measuring volumes with a diameter of 125 mm and a height of 150 mm may be irradiated by a radiation output of 39 W and a voltage of 130 kVolt. CT measurements can be used to analyse defects as pores, blowholes, inclusions and to and to evaluate the porosity. Initially the mentioned parts get treated with X-Rays and several hundreds of radiographs with full rotation of the components are acquired. Subsequently through the software VG-StudioMAX a 3D Model is reconstructed that acts as evaluation. The results of the CTanalysis are shown in Table 6.6 and 6.7. The untreated specimens have a percentage of porosity of around 3.05 % whereas the HIPed specimens only have 0.81 %. So the percentage of porosity could be reduced by around 2.2 %. Also the volume of the biggest pore could be reduced from 0.024 mm^3 to 0.009 mm^3 by around 160 %, but the numbers of biggest pores turned out to increase. Before the HIP-process in average around 21 pores had a size nearby the maximum (0.033 mm^3). After HIP around 49 pores were detected. The reason is that HIP compressed the large pores whereas the amount of small pores rose.

Table 6.6: CT Results of Non-HIP ed Samples [12]

No	Percentage of Porsity (%)	Vol. of the biggest pore V(mm³)	Quantity of detected pores n
1	2.84	0.01	24.00
2	3.08	0.023	19.00
3	3.42	0.030	31.00
4	2.84	0.033	11.00
Average	3.05	0.024	21.25

The average porosity of an untreated specimen is around 3.05 % which means 96.95 % density. For a HIPed specimen the porosity is around 0.81 % which means 99.19 % density. The comparison with the Archimedean density of 98.14 % shows a deviation between the methods due to the lack of validation.

Table 6.7: CT Results of HIP ed Samples [12]

No	Percentage of Porsity (%)	Vol. of the biggest pore V(mm³)	Quantity of detected pores n
1	0.77	0.009	61.00
2	0.85	0.006	67.00
3	0.67	0.010	10.00
4	0.94	0.012	57.00
Average	0.81	0.009	48.75

References

[1] Barksdale, Jelks, "Titanium". In Clifford A. Hampel (ed.). *"The Encyclopedia of the Chemical Elements"*, New York: Reinhold Book Corporation. 1968, pg. 732–738. LCCN 68029938.

[2] Yang, X., & Richard Liu, C, "MACHINING TITANIUM AND ITS ALLOYS", *Machining Science and Technology*, 3(1), 1999, pg. 107-139.

[3] Donachie, Matthew J., Jr. (1988). *TITANIUM: A Technical Guide. Metals Park*, OH: ASM International. p. 11. ISBN 978-0-87170-309-5.

[4] Lide, D.R., ed, *"CRC Handbook of Chemistry and Physics"* (86th ed.), 2005, Boca Raton (FL): CRC Press. ISBN 0-8493-0486-5.

[5] https://matmatch.com/learn/material/titanium

[6] https://www.dorsetware.com/steel-vs-titanium/

[7] https://www.azom.com/article.aspx?ArticleID=1336

[8] https://www.sharrettsplating.com/base-materials/refractory-metals

[9] https://www.lenntech.com/periodic/elements/ti.htm

[10] Gu, D., Hagedorn, Y.-C., Meiners, W., Meng, G., Batista, R. J. S., Wissenbach, K., & Poprawe, R, "Densification behavior, microstructure evolution, and wear performance of selective laser melting processed commercially pure titanium", *Acta Materialia,* 60(9), 2012, pg. 3849–3860.

[11] Gu, D., & Shen, Y, "Effects of processing parameters on consolidation and microstructure of W–Cu components by DMLS", *Journal of Alloys and Compounds,* 473(1-2), 2009, pg. 107–115.

[12] Uhlmann, E., Kersting, R., Klein, T.B., Cruz, M.F., & Borille, A.V, "Additive Manufacturing of Titanium Alloy for Aircraft Components", *Procedia CIRP,* 35, 2015, pg. 55–60.

[13] DIN5832-2. *Implants for surgery - Metallic materials* - Part 2: Unalloyed titanium (ISO 5832-2:1999); German version EN ISO 5832-2:2012.

[14] DIN EN ISO 3252: *Powder metallurgy; Berlin: Beuth,* February 2001.

[15] DIN66165-2. *Particle size analysis; Sieve Analysis - Implementation;* Berlin: Beuth, April 1987.

[16] DIN EN ISO 3923-1: *Metal powder - Determination of bulk density - Part 1:* Funnel method; Berlin: Beuth, April 2010.

[17] DIN 17869. *Material properties of titanium and titanium alloys - Additional information;* Berlin: Beuth, June 1992.

[18] DIN 65083 (Design). *Aerospace series - Heat treatment of titanium and titanium alloy castings;* Berlin: Beuth, Nov. 2011.

[19] DIN 65083 (Design). *Aerospace series - Heat treatment of titanium and titanium alloy castings;* Berlin: Beuth, Nov. 2011.

[20] Bodycote: Hot isostatic pressing; Menden; Brochure; April 2010.

[21] DIN ISO 3369. Impermeable sintered metal materials and hardmetals - Determination of density; Berlin: Beuth 2006.

[22] Sartorius AG: *Weighing Applications Manual - Diche; Part 1: Sartorius; Marketing weighing technology;* Goettingen. Corporate font, Feb 2001.

[23] Mercelis P, Krut h JP, "Residual stresses in selective laser sintering and selective laser melting", *Rapid Prototyping J,* 12(5); 2006, pg. 254-265.

[24] Sartorius AG: *Weighing Applications Manual – Diche; Part 1: Sartorius; Marketing weighing technology;* Goettingen. Corporate font, Feb 2001.

[25] DIN 6507-1. *Metallic materials - Vickers hardness test* - Part 1: Test method (ISO 6507-1:2005); German version EN ISO 6507-1:2005.

[26] Matucha KH, *"Structure and properties of nonferrous alloys"*, Weinheim: VCH; 1996.

[27] Mario Viespoli, L., Bressan, S., Itoh, T., Hiyoshi, N., Gokuldoss Prashanth, K., & Berto, F, "Creep and high temperature fatigue performance of as build selective laser melted Ti-based 6Al-4V titanium alloy", *Engineering Failure Analysis,* 2020, 104477.

[28] M. Wu, T. Itoh, Y. Shimizu, H. Nakamura, M. Takanashi, "Low cycle fatigue life of Ti–6Al–4V alloy under non-proportional loading", *Int. J. Fatigue,* 44, 2012, pg. 14–20.

[29] Rafi, H.K., Karthik, N.V., Gong, H., Starr, T.L., & Stucker, B.E, "Microstructures and Mechanical Properties of Ti6Al4V Parts Fabricated by Selective Laser Melting and Electron Beam Melting." *Journal of Materials Engineering and Performance,* 22(12), 2013, pg. 3872–3883.

[30] E. Brinksmeier, G. Levy, D. Meyer, and A.B. Spierings, "Surface Integrity of Selective-Laser-Melted Components", *CIRP Ann. Manuf. Technol.,* 59(1), 2010, pg. 601–606

[31] J.P. Kruth, G. Levy, F. Klocke, and T.H.C. Childs, "Consolidation Phenomena in Laser and Powder-Bed Based Layered Manufacturing", *CIRP Ann. Manuf. Technol.,* 56, 2007, pg. 730–759

[32] Leuders, S., Thöne, M., Riemer, A., Niendorf, T., Tröster, T., Richard, H.A., & Maier, H.J, "On the mechanical behaviour of titanium alloy TiAl6V4 manufactured by selective laser melting: Fatigue resistance and crack growth performance", *International Journal of Fatigue,* 48, 2013, pg. 300–307.

[33] Hück, M. (1983). An improved method for evaluating stair step tests. *Materials Science and Technology, 14*(12), 406-417.

[34] Dixon W.J., & Mood AM, "A method for obtaining and analyzing sensitivity data", *J Am Stat Assoc,* 43, 1948, pg. 109–26.

[35] L. Badea, M. Surand, J. Ruau, B. Viguier, "Creep behavior of Ti-6Al-4V from 450°C to 600°C", Univ. *Polytechnica of Bucharest Sci. Bull., Seri. B* 76, 2014, pg. 185–196.

[36] L.A.N.S. Briguente, A.A. Couto, N.M. Guimaraes, D.A.P. Reis, C. Moura Neto, M.J.R. Barboza, "Determination of Creep Parameters of Ti-6Al-4V with bimodal and equiaxed microstructure", *Defect Diff. Forum* 326–328, 2012, pg. 520–524.

[37] E. Brinksmeier, G. Levy, D. Meyer, and A.B. Spierings, "Surface Integrity of Selective-Laser-Melted Components", *CIRP Ann. Manuf. Technol.,* 59(1), 2010, pg. 601–606

[38] J.P. Kruth, G. Levy, F. Klocke, and T.H.C. Childs, "Consolidation Phenomena in Laser and Powder-Bed Based Layered Manufacturing", *CIRP Ann. Manuf. Technol.,* 56, 2007, pg. 730–759

[39] ASM Handbook, Vol. 1, ASM International, *Materials Park, OH,* 1993, p 2071.

[40] T. Vilaro, C. Colin, and J.D. Bartout, "As-Fabricated and Heat Treated Microstructures of the Ti-6Al-4V Alloy Processed by Selective Laser Melting", *Metall. Mater. Trans. A.,* 2011, 42, pg. 3190

[41] L. Facchini, E. Magalini, P. Robotti, A. Molinari, S. Hogess, and K. Wissenbach, "Ductility of Ti-6Al-4V Alloy Produced by Selective Laser Melting of Pre-Alloyed Powders", *Rapid Prototyp. J,* 2010, 16(6), pg. 450–459.

[42] B. Vrancken, L. Thijis, J.P. Kruth, and J.V. Humbeeck, "Heat Treatment of Ti6Al4V Produced by Selective Laser Melting—Microstructure and Mechanical Properties", *J. Alloy Compd.,* 2012, 541, p 177–185.

[43] Blumenauer, H., & Pusch, G. (1993). *Technical fracture mechanics German publisher for basic industry.* Leipzig, Stuttgart. [in German].

Chapter VII

Selective Laser Melting (SLM)

of

Inconel

(Ni based Alloys)

CHAPTER VII

7. Selective Laser Melting (SLM) of Inconel (Ni based Alloy)

7.1. Introduction

Inconel is a family of high-performance nickel-based alloys developed by Special Metals Corporation. It is one of the most widely recognised trademarks in the industry, associated with the original development of 18 different alloys [1]. The one thing that each Inconel alloy has in common is that chromium is the main alloying element after nickel, although a far wider range of additional elements are used to provide varying levels of properties and performance.

The very first Inconel alloy brought to market was Inconel 600. It now finds use in nuclear reactors, but the original target application was far humbler: milk cans. Based upon a chemical composition of >72% nickel, with up to 17% chromium and less than 10% iron, it was one of the first alloys to perform well at elevated temperatures of up to 600°C.

The latest alloy to carry the Special Metals Corporation trademark is Inconel 693. This alloy was launched in 2006 and was developed specifically to resist metal dusting a phenomenon that can occur in gas synthesis reactors. It can resist oxidation and carburisation up to as high as 1150°C, partly due to its very high chromium content of up to 31%.

Nickel-based alloys offer good performance at higher temperatures. The melting temperature of nickel is 1453°C, which is far higher than metals such as copper (1084°C) and aluminium (660°C). The ability to operate at elevated temperatures though is not only linked to metals melting temperature, otherwise iron (1150°C) or steel (1400°C) would be more widely used. The additional property of nickel and nickel alloys which allows them to operate at elevated temperatures is their ability to form a thick, stable, passive oxide layer when heated which protects them from further attack. This oxide layer can be several microns in thickness, dependent upon the temperature and environment to which the metal has been exposed. Perhaps the most widely specified alloys, particularly in Oil and Gas related applications, are Inconel 718 and Inconel 625.

Inconel 718 was launched in 1962, initially for the aerospace industry, but due to its ready availability in the market it was taken-up by the oil and gas industry. This was associated with wells moving offshore, drilling deeper with higher temperatures, pressures and corrosive contaminants. Inconel 718 utilises precipitation treatment to achieve significantly increased strengths, whilst still retaining much of the toughness (impact strength) and formability (elongation) of the base alloy.

Inconel 625 was launched just after Inconel 718 in 1964. Inconel 625 achieves reasonably high strength through the addition of molybdenum and niobium to the nickel-chromium base, but nowhere as high as Inconel 718. However, due to the high level of chromium and molybdenum it is corrosion resistant in the most aggressive of environments. It is particularly resistant to pitting and crevice corrosion with a pitting resistance equivalent number (PREN) of over 45.

Langley Alloys stocks Alloy 718 (2.4668, UNS N07718) and Alloy 625 (2.4856, UNS N06625) as solid bars between 5/8" (15.875mm) and 10" (254mm) diameter. Our stock is certified to API 6A CRA and NACE MR0175 / ISO 15156-3. Inconel 725 was developed in 1989 and seeks to address the obvious limitation of Inconel 625 by increasing its strength. Retaining the same basic composition as Inconel 625, and therefore the same excellent corrosion resistance, additions of titanium are used in combination with further heat treatment to achieve age-hardening.

Langley Alloys stocks Alloy 725 (UNS N07725) as solid bars between 1" (25.4mm) and 4" (101.6mm) diameter, suitable for fastener manufacturer. Our stock is certified to API 6A CRA and NACE MR0175 / ISO 15156-3. Super alloys are high performance alloys have become the steel of choice for corrosion resistance and versatility [2].

7.1.1. Characteristics of Inconel

- Good resistance to acids, such as sulphuric, phosphoric, nitric, and hydrochloric.
- Almost completely free from chloride induced stress corrosion cracking.
- Excellent mechanical properties at both extremely low and extremely high temperatures.
- Outstanding resistance to pitting, crevice corrosion and intercrystalline corrosion.
- High resistance to oxidation at elevated temperatures.

7.2. Applications of Inconel

- Chemical and petrochemical processing.
- Components where exposure to high mechanical stress and sea water are required.
- Flare stacks on offshore oil platforms.
- Gas turbines, rocket motors/engines, and space craft.
- Pollution control equipment.
- Nuclear reactors.

7.2.1. *Oil & Gas Extraction*

Inconel is ideally used in the oil and gas extraction industries due to its high temperature resistance and oxidation resistant properties. The oil and gas industries need a super alloy metal, like Inconel, that can withstand extreme environments and volatile, corrosive gasses. The super alloy Inconel 625 is especially useful for the processing systems required for natural gas production. Due to Inconel 625's particularly strong thermal fatigue strength and oxidation resistance, it's often used for the separation of extracted fluids or in line steel transfer piping.

7.2.2. *Heat Treatment Applications*

Inconel is famously resistant to extreme temperatures, and retains enough tensile strength at high temperatures to continue holding moderate loads (Inconel 625 retains 13.3 ksi, tensile strength at 2,000°F). This makes Inconel the ideal basket material for heat treat applications—comparing favourably to stainless steel alloys such as Grade 304, 316, and 330 SS. Compared to most stainless steel alloys, a basket made from a superalloy like Inconel won't lose shape as easily when holding parts through a rigorous heat treat application.

7.2.3. *Rapid Temperature Changes*

Some manufacturing processes may combine high and low temperature processes in rapid succession. Most Inconel alloys retain excellent oxidation resistance at high and low temperatures, allowing a single basket made from Inconel to be used in processes where temperatures vary between near-cryogenic lows and heat treatment highs.

7.2.4. *Saltwater Applications*

Inconel is often used in marine applications because of its extraordinary resistance to sodium chloride (salt) at a variety of temperatures. So, for processes that use salt or factory locations near the ocean, Inconel can be ideal for a parts washing basket. However, some stainless steel alloys also exhibit excellent resistance to saltwater. So, when would Inconel is considered preferable? Generally speaking, this super alloy would only be much more useful than the grade 316 stainless steel, if extreme temperatures in excess of 1,000°F would be a concern as well. Inconel would retain its oxidation resistance better at high temperatures than 316 SS would.

7.2.5. *Jet Engines*

Some of Marlin's manufacturing clients might use Inconel in the construction of jet engines and turbines. Jet engines are unique in that they have to withstand extreme temperatures both high and low frequently during use. As noted on the meteorology training website, at 36,000 ft.

(well within a passenger jet's standard cruising altitude), the average air temperature is -56.3°C (-69.3°F), but the combustion process can exceed temperatures of 1,150°C (2120°F).

Cooling technologies are used to rapidly reduce temperatures, but they can still easily exceed the tolerances of many metals. Inconel 600 retains high oxidation resistance and tensile strength despite rapid changes in temperature from the combustion process and cooling technologies employed in the aerospace industry. Basically, Inconel is an ideal material whenever extreme temperature and chemical resistance are a must, and for any process where temperature highs would normally degrade the oxidation resistance of other metals.

7.3. Densification behaviour based on Laser Related Process Factors

Choi, Joon-Phil et al [3] investigated the densification behaviour and microstructural development of Inconel 718 parts fabricated by selective laser melting (SLM) were investigated with variation of the applied laser scanning speed. Materials of Inconel 718 powder particles were prepared using a gas atomization process showing uniform size distribution of 10-45 μm with high flowability. IN718 samples were fabricated by SLM at various scanning speeds and the densification and resultant characteristics of the microstructure were investigated.

The SLM process was carried out using a Concept Laser Mlab-Cusing system, equipped with a 100 W continuous wave (CW) Yb:YAG laser with a beam diameter of 110 μm and wavelength of 1075 nm. The IN718 cuboid specimens (10 × 10 × 10 mm, Figure 7.1) were manufactured as shown in the figure 1 using laser power (P) of90 W, layer thickness (t) of 25 μm, hatch spacing (h) of 80 μm and laser scanning speeds (v) of between 100 and 1600 mm/s. All specimens were prepared using a Z-increment under a high purity Argon atmosphere containing<0.3% oxygen, and the continuous line scanning strategy (zigzag) was adopted in this work.

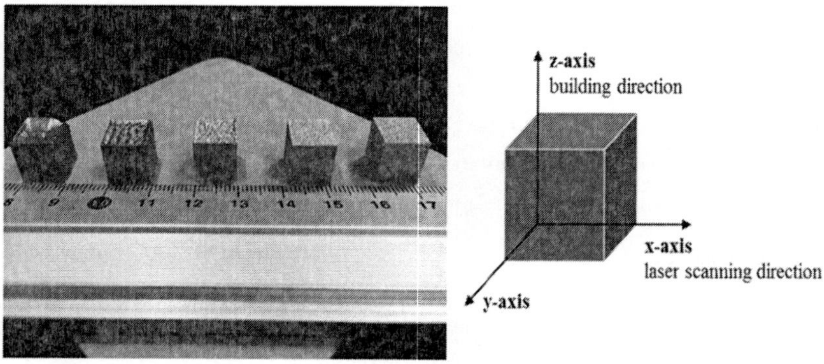

Figure 7.1: Inconel 718 Samples Built with different Laser Scanning Speeds [3]

As seen in figure 7.2, the gas-atomizedIN718 powder particles were generally spherical in shape with limited amount of irregular morphologies, and smaller satellite particles attached to the main particles were clearly visible. By sectioning the powder particles as shown in figure 7.2(b), it was discovered that several pores had been generated within a few particles, probably by gas entrapment during the gas atomization.

The size distribution of the IN718 powders revealed a narrow particle size distribution with a mean diameter(D_{50}) of 34 μm as shown in figure 7.2(c), as measured using a laser diffraction particle size analyser (Beckman Coulter LS 13 320, US). The flow characteristics of the IN718 powder are summarized in Table 7.1.

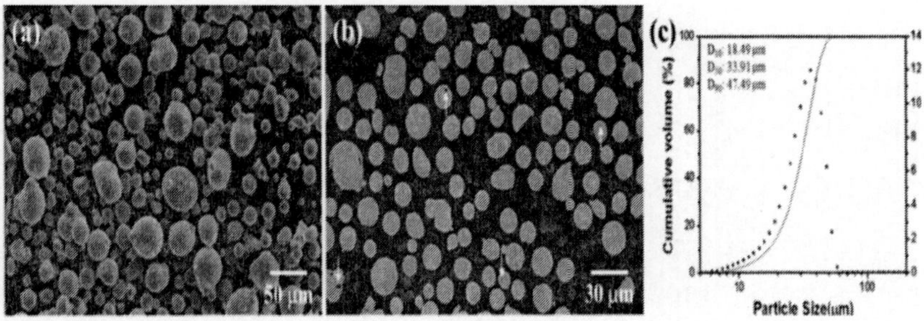

Figure 7.2: SEM Micrographs of (a) As-received Inconel 718 Powder Particles, (b) Their cross Sections, with White Arrows Indicating Pores in the Particles, and (c) Particle Size Distribution [3]

Table 7.1: Flow Characteristics of IN728 [3]

True density g/cm³	**Hall flow Sec/50g**	**Apparent density g/cm³**	**Tapped density g/cm³**	**Hausner ratio**
8.19±0.01	20.20± 0.20	4.19 ± 0.02	5.09±0.05	1.21

The powder flowability was tested with a hall flow meter according to ASTM B213-03, and showed 20.2 ± 0.2 s for 50 g IN718 powder, indicating good flowability [4, 5]. The apparent density and tapped density were also evaluated (4.19 g/cm³ and 5.09 g/cm³, respectively) using a hall flow meter following ASTM B212-09 [5]. The Hausner ratio is correlated with the powder flowability, defined as the ratio between the tapped density and apparent density. Generally, the values < 1.25are considered to indicate high flowability. In this work, the Hausner ratio of IN718 powder was 1.21, suggesting that the powder had good flow properties.

Figure 7.3 depicts the effect of laser scanning speed on the porosity of the SLMed samples and representative optical microstructures; while the laser powder, laser thickness, and hatch spacing were maintained at constant value of 90 W, 25 μm, and 80 μm, respectively. As seen in figure 7.3(a), the porosity of as-built samples was significantly influenced by scanning speed.

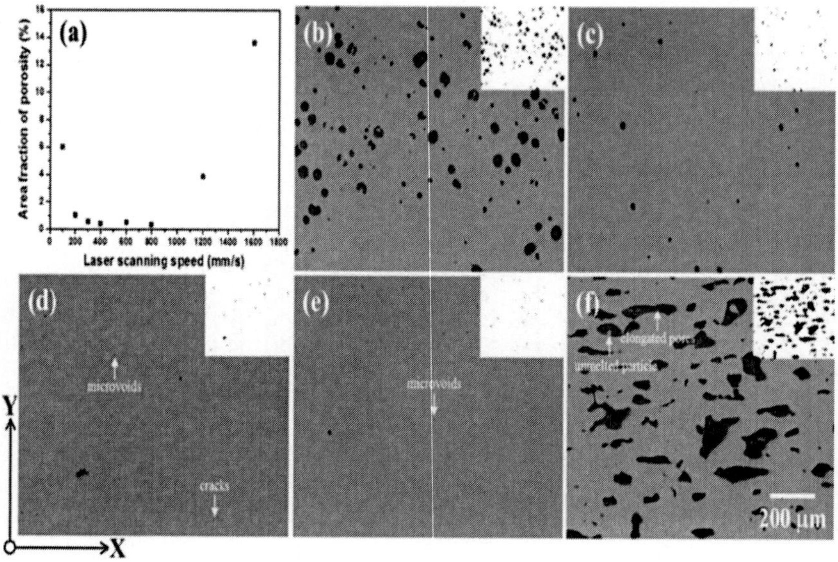

Figure 7.3: (a) Variation of Porosity of SLM IN718 samples with different Laser Scanning Speeds and (b–f) Corresponding OM Images of Selected Samples: (b) 100, (c) 200, (d) 400, (e) 800, and (f) 1600 mm/s [3]

The porosities decreased with decreasing the scanning speed from 1600 mm/s to 800 mm/s, and low porosity levels < 1% were maintained at laser scanning speeds of 200 to 800 mm/s. At this point, the lowest porosity of 0.35%was found with as-built IN718 samples using a laser scanning speed of 800 mm/s. This densification behaviour is mainly attributed to the variation of applied laser energy during SLM. For given materials, the energy density E (in J/mm^3) is defined as the applied laser energy per unit volume, and this can be considered as a critical factor that affects the quality of SLM parts, as shown in Eq. (1):

$$E = \frac{P}{vht} \qquad (7.1)$$

Where, P is the laser power (W), v is the scanning speed (mm/s), h is the hatch spacing (mm), and t is the layer thickness. Thus, in this study, it was simple to control the laser energy density by managing the scanning speed. In general, a higher incident laser energy density induced by low scanning speed leads to a large amount of melting that generates a fully molten pool. This condition increases the final density [6, 7].

However, there is no direct relationship between relative density and applied energy density as described in figure 3(a), because the porosity of the as-built samples increase at very low scan speeds (< 200 mm/s). This can be attributed to the balling and dross formation that occurred in the melt pool, resulting in a poor surface finish and lower density [8].The microstructures of cross-sections produced at different scanning speeds were compared in figure 7.3(b)–(f), and the results were consistent with the porosity values reported in figure 7.3(a). At a low scanning speed of 100 mm/s (figure 7.3(b)); large spherical pores were located over the entire surface of the as-built IN718 specimen. This scanning speed gave rise to excessive energy input (E = 450 J/mm^3)during the SLM process and produced over-melting on the powder bed, which can cause vaporization of powder materials and gases trapped within the melt pool. This leads to near-spherical defects during solidification [9, 10]. As the scanning speed increased from 100 to 200 mm/s, shown in figure 7.3(c), the porosity decreased from 8.24 to 1.05%, but a few observable pores were present in the SLM samples. On raising the laser scanning speed to 400 mm/s, and then 800 mm/s, as depicted in figure 7.3(d) and (e); the porosity was calculated to be 0.43 and 0.35%, respectively. This indicated full densification by elimination of the imperfections and large pores but a small number of cracks and micro voids were still detectable in the microstructure.

On the other hand, on increasing the laser scanning speed to 1600mm/s as seen in figure 7.3(f), discontinuous laser scan tracks were observed throughout the cross-section, along with a large number of irregularly shaped pores, and also the elongated pores formed along the laser scanning direction.

Based on the above experimental results, it is pertinent to consider that the densification behaviour of SLM IN718 parts is determined by the laser scanning speed, and that near-full density parts (> 99%) could be obtained at a laser power of 90 W with a scan speed range of 300–800 mm/s, by minimizing macro defects. This zone, therefore, indicates a stable processing condition for the current investigation, corresponding to input laser energy density of 60–150 J/mm^3.

Furthermore, the melt pool size as marked by the white line in figure 7.4(a) was determined to be approximately 75 μm in all specimens, causing about 30% overlap of the melt pool width between neighbouring scan tracks. This is in good agreement with the applied hatch spacing of

80 μm. In the SLM sample fabricated at 200 mm/s, there were some large pores located in the melt pool indicated in figure 7.4(a), whereas in the samples fabricated at 400 and 800 mm/s shown in (figure 7.4(b) and (c), no obvious macro-pores were observed at the surface of the samples. This is consistent with the porosity earlier observations were also discussed with respect to figure 7.3.

7.4. Sustainability of Ni based Alloy

Ardila, L. C., et al [16] and his team investigated the effect of IN718 Recycled Powder Reuse on Properties of Parts Manufactured by Means of Selective Laser Melting along with the processes, experimentally carried out by using an industrial SLM system developed by MCP (Model Realizer SLM 250).

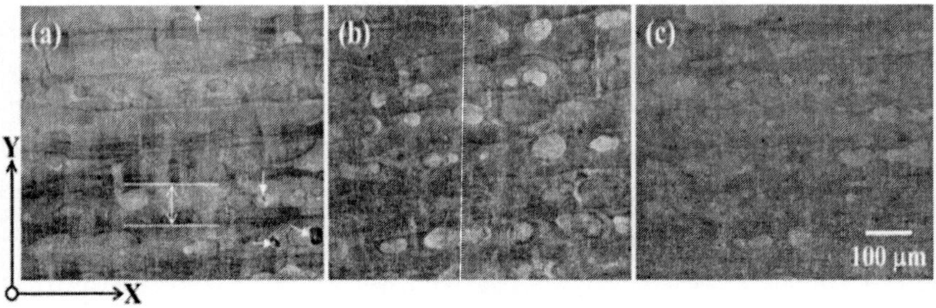

Figure 7.4: Optical Micrographs for Cross-sections of SLM Samples Fabricated at different Scanning Speeds (a) 200, (b) 400, and (c) 800 mm/s [3]

The device is equipped with a200 W fiber laser and works in a protective argon atmosphere with an oxygen content limited to a 0.2 %. The powder used for the experiment was a single 25 kg batch of Inconel 718 Ni super alloy, obtained by atomization with argon gas as this is the most suitable method to obtain high quality powders with small particle size distribution and large homogeneity. The material was produced by LPW Tech. Ltd., with the chemical compositions indicated in Table 7.2 (chemistry to AMS 5832 F) and with a particle size in the range 15-45 micron, according to material certification.

To study the effect of powder reutilization after recycling, a very specific methodology was developed. It consists of an iterative process distributed in three main steps as shown in figure 5. The first one, after all the powder is loaded into the SLM device and proper fabrication parameters are set, is the fabrication of test samples. These samples are removed from the machine and stored apart for a posterior metallurgical and mechanical analysis. The second step requires removing of all powder from the fabrication chamber to submit it to consecutive sieving and drying processes. In the third and last step, part of the powder is separated for the study of its properties, and the rest is loaded into the SLM device to start over the fabrication process again.

Table 7.2: Chemical Composition of the IN718 Powder [11]

Alloy	% Ni	% Cr	% Co	% C	% Mo	% Al	%Ti	% Fe	%Nb +Ta	% Mn
IN718	Bal	18.64	< 0.3	0.06	3.05	0.62	0.93	17.53	4.88	0.03

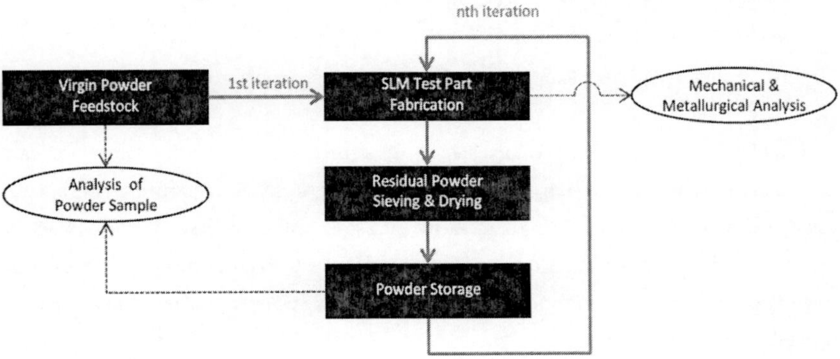

Figure 7.5: Powder Sampling Methodology [11]

By doing this, IN718 powder is expected to recover its initial properties at the beginning of the following iteration and therefore, to allow for fabrication of parts with identical mechanical properties, thus optimizing resource utilization by minimization of wastes. To check the validity of this approach, metallurgical and mechanical analysis of samples was accomplished via measurements of: porosity, using an Olympus GX51 optical microscope; material toughness through Charpy impact tests performed with a Hoytom 300J/AD2 system; material hardness with a DuraScan microdurometer from EmcoTest; and microstructure using an UltraPlus SEM system from Zeis, which in addition incorporates EDS and EBSD capabilities and allowed for a complete powder composition characterization.

Fourteen loops of this methodology were executed using an optimized set of parameters (for IN718) and a layer thickness of 20 μm to ensure the best possible quality of test samples. At every iteration, six mechanical test samples with dimensions of 60x12x12 mm and three metallurgical test parts (dimensions: 10x10x10 mm) were built in the industrial SLM system to later perform a metallurgical and mechanical analysis and thus study the dependence of part properties with powder ageing as shown in figure 7.6.

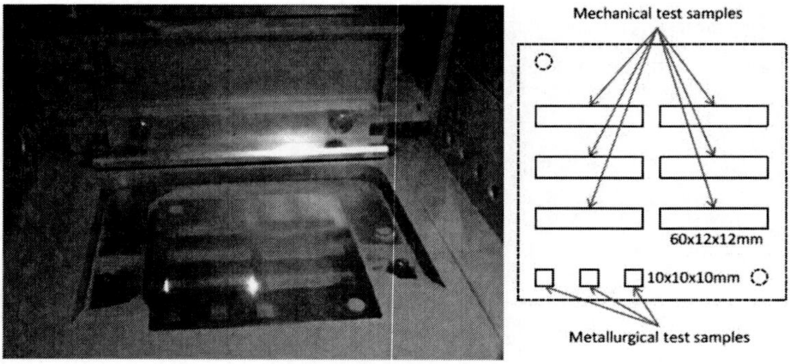

Figure 7.6: Picture Showing Manufacturing of Tests Pieces; Arrangement and Dimensions of Test Pieces [11]

In addition to this, powder volumes (150 g) were collected in every loop after each corresponding sieving and drying processes for their analysis as well. In particular, the powder was sieved at 20-25 ºC to remove agglomerates using standard stainless steel sieve with 63 μm mesh size and then dried in an oven with air circulation to remove moisture. Trials were properly documented and best practices in powder handling were carried out during the work to avoid any material pollution.

The analysis of powder samples was focused on the two main parameters that allow their characterization: particle size distribution and material composition. For the first one, scanning electron microscopy (SEM) was employed. Examples have been depicted in Figure 7.7(a) and 7(b) for the initial (n = 1) and last iterations (n = 14), where it can be easily observed the appearance of particle aggregates in the second one due to the sintering of powder beads during manufacturing.

This is confirmed in Figure 7.7(c), which shows the particle size distribution obtained for three different iterations (n= 1, 7, and 14). As one can see, the distribution remains almost identical from iterations 1 to 7, while there is a small deviation from iterations 7 to 14. This result ratifies that a continuous use of the same IN718 powder, iteration after iteration, gives rise to undesired particle aggregations which may affect the quality of the fabricated part. However, the modification of this distribution is not very significant thanks to the methodology employed; the variation with respect to the nominal values is less than 10%, and consequently test parts fabricated with aged powder are not expected to show relevant quality differences with respect to those fabricated with the initial unused powder. In order to prove this assumption, metallurgical and mechanical measurements were additionally performed and properly analysed in the following sections.

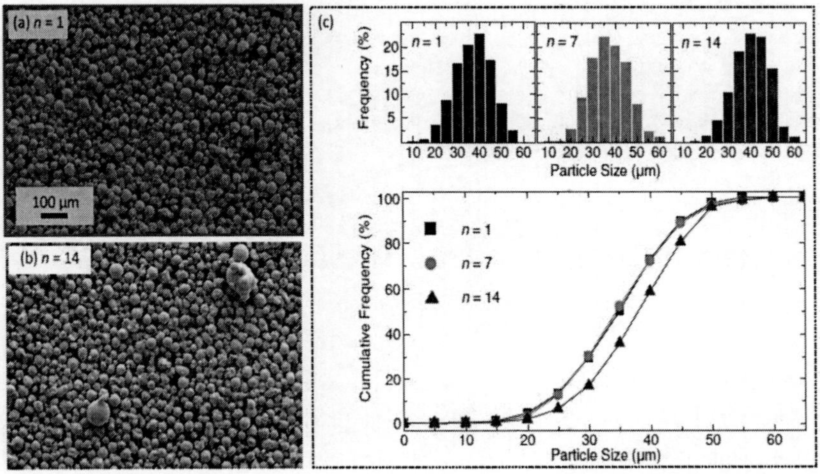

Figure 7.7: SEM Images of Powder (a) before and (b) after 14 Iterations Sieving and Drying. Comparison of Particle Size Distribution is shown in (c) in Terms of Relative (top) and Cumulative (bottom) Frequencies [11]

The study of the material composition dependence on IN718 powder reutilization was carried out by obtaining energy-dispersive X-ray spectroscopy (EDS) measurements. Figures 7.8(a) & (b) respectively shows the spectral powder composition for the first and last iterations. As it can be observed, such composition remains virtually the same with the increased number of iterations.

The variation of composition relies within the standard deviation values for all elements, except for Ni and Nb where this variation is larger than expected. The concentration of Ni is smaller at the final iteration while much bigger for Nb (and moderately for other elements like Cr and Fe).

This result may be originated by a slight oxidation of Ni as a function of subsequent fabrication processes, but overall IN718 powder is not oxidized, and therefore one may assume that test part properties will remain virtually identical and irrespective of the fabrication process iteration.

To prove this assumption, a complete metallurgical characterization of test parts was performed by means of porosity and microstructural measurements. Porosity was measured by analyzing material density in the inner region of the fabricated parts with an optical microscope.

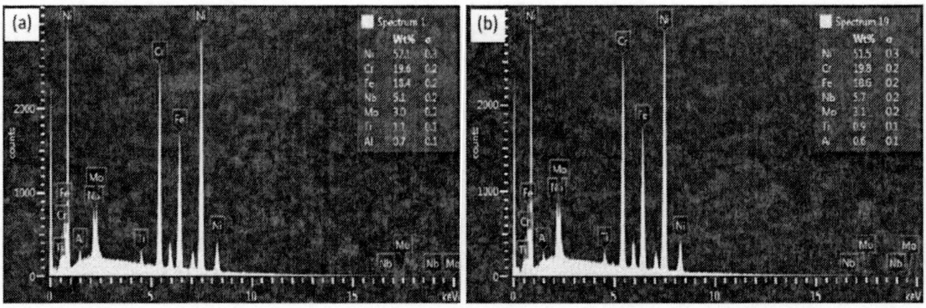

Figure 7.8: EDS Spectra of IN 718 Powder Composition for the (a) First and (b) Last Iterations [11]

The inner part of the metallurgical test samples (cubes from figure 6) was reached by cutting them in half with a high precision saw from Bhueler (model Isomet 4000), and then polished with a Mecatech 334polisher form Presi. Afterwards, porosity is obtained by comparing 10 representative photos from each sample. These photos are analysed with specific software analysis included in the microscope which allows distinguishing the pores from the base material by means of correcting contrast and brightness. The programme gives the pore percentage present in each micrograph. The average of the ten micrographs is the value of the sample's porosity.

Due to the large amount of test parts fabricated in this study, only those corresponding to five particular iterations were analysed (n = 1, 4, 7, 10, and 14). In each one of these iterations, three metallurgical test samples were fabricated, as mentioned before. After doing porosity measurements over them, average values and standard deviations were calculated. The results have been presented in figure 7.9. The dashed line represents the overall average porosity value. As it can be observed, porosity is very similar in all iterations, close to a 0.10%, while standard deviations show more variations but not in a relevant way. These results points out to the same conclusions obtained in the previous sub-section, that the recycling process allows the fabrication of parts with reused powder with similar properties.

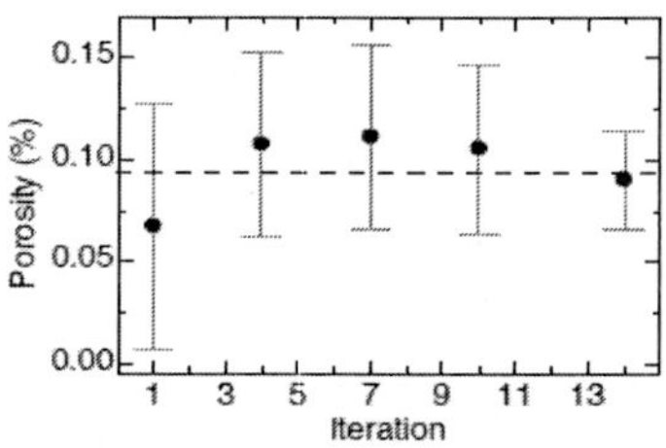

Figure 7.9: Porosity Measurements Performed in Manufactured Samples [11]

Microstructure of IN718 test samples was revealed using Kalling's Reagent. Figures 7.10(a) and 7.10(b) show examples of the microstructure found in the cross and longitudinal respectively. A fine dendritic microstructure is formed due to very high cooling rate induced by high laser energy proper of this kind of technology.

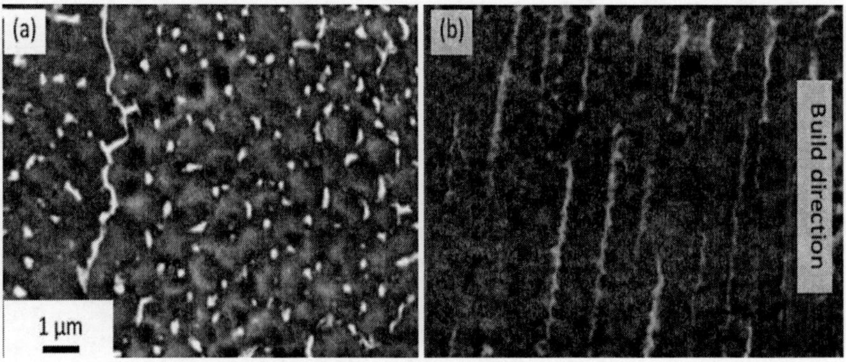

Fig. 7.10: SEM Micrographs of IN718 (a) SLM Manufactured Samples on Cross Section, (b) Longitudinal Section [11]

The cross section micrograph in figure 7.10(a) shows that the dendritic structure is not homogeneous due to the interaction of melted tracks under stripped pattern manufacturing strategy (X/Y cross-hatching technique).

In contrast, figure 7.10(b) shows a regular solidification microstructure along longitudinal direction is clearly formed, which is composed of parallel dendrites whose growing parallel to Z-direction. In SLM process, cooling solidification of the liquid metal is rapidly produced due to high dissipation of heat through base plate; which creates high temperature gradients and solidification rate in the Z-direction, leading to formation of columnar dendrite grains in Z direction. At higher magnifications seen figure 7.10(a) different gamma and Laves phases can be found (an embrittling phase, such as [Ti, Nb] C type carbides).

In total accordance with the results found in the previous EDS spectra, the observed microstructure is repeated in all tests samples, irrespectively of the iteration in which they were fabricated as shown in figure 7.11. Basically no differences are observed in terms of precipitates and phases, which is consistent with the reutilization of powder presenting nearly identical properties in terms of size and composition.

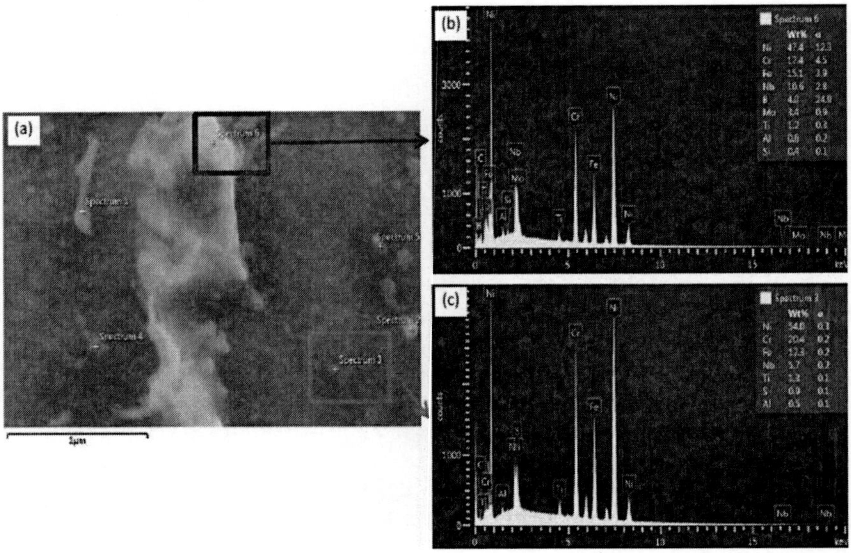

Figure 7.11: (a) Matrix γ and Phase Laves, (b and c) Corresponding EDS Analysis [11]

7.5. Mechanical Properties of SLMed Parts

7.5.1. *Hardness*

Choi, Joon-Phil [3] investigated the densification behaviour along with the mechanical properties of Inconel 718 parts fabricated by selective laser melting (SLM) were investigated with variation of the applied laser scanning speed as discussed in section 7.3. The Vickers hardness of the SLM samples fabricated at different laser scanning speeds is illustrated in figure 7.12.

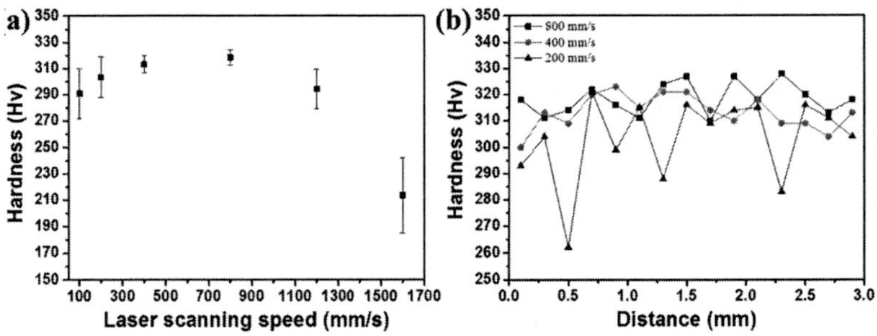

Figure 7.12: Vickers Hardness in the XY Plane of SLM IN718 Samples with different Laser Scanning Speeds: (a) Average Hardness Value and (b) Hardness Distribution [3]

It was found that the hardness value varied according to the laser scanning speed, and that there was a correlation between the hardness and porosity of a part by comparing figure 7.12(a) and 7.12(b): the hardness decreased as porosity increased. This is mainly because the pores within these materials, especially irregular shaped and relatively large pores, collapsed easily under load during the Vickers hardness test [12].

Figure 7.12(b) presents the hardness profile of SLM IN718 samples with near full density (> 99%). The average hardness of the dense samples was 311.7 $H_{v0.5}$, and the maximum hardness value (318.5 $H_{v0.5}$) was obtained at the laser scanning speed of 800 mm/s. This result was comparable or exceeded that of conventionally manufactured parts [13, 14]. However, the dense samples exhibited a similar hardness value, but with a very different standard deviation. When the laser scanning speed was varied from 800 to 200 mm/s, the hardness decreased to 303.3 Hv0.5, meanwhile, the fluctuation of hardness values intensified (± 15.5 Hv0.5). This large fluctuation in the hardness profile is believed to be caused by in homogeneously distributed microstructural features, which may have negative effects on the mechanical properties of the SLM components [15, 16]. It is therefore apparent that the optimum laser scanning speed in this study was 800 mm/s, in order to fabricate near full density parts with sound microstructure and acceptable mechanical properties.

7.5.2. Impact Strength

Ardila, L. C., et al [17] and his team investigated the effect of IN718 Recycled Powder Reuse on Properties of Parts by charpy test; this type of test was selected in order to analyse the sensitivity of the SLM process to powder degradation. In this compared to other tests, like tensile test, any little difference in mechanical properties results in a big change in the measured values, by Schierra. Charpy test scans the material around the notch and the weakest link will be found, thus reflecting weakness of the microstructure, with resulting energy of the test. To do so, standard size and V-shaped notch specimens were machined and tested according to ISO 148-1 standard. Prior to testing, all test samples were heat treated following AMS 5662 standard for IN718 which is solution treatment and double aged and the results were shown in figure 7.13.

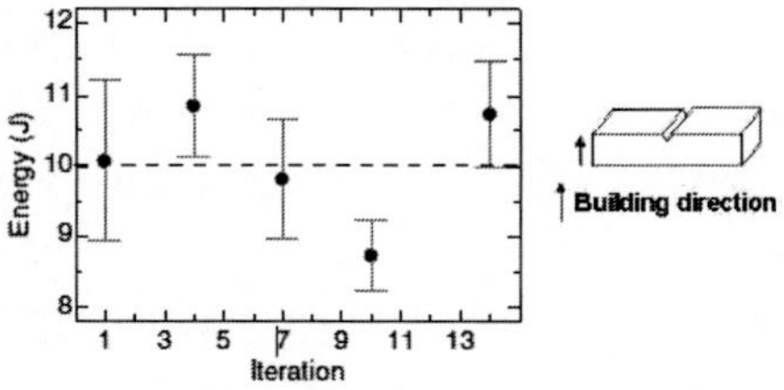

Figure 7.13: Charpy Impact Tests Performed over Manufactured Samples [17]

It is worth to note here that the building direction may play an important role in these tests, since mechanical properties could depend on part axis direction. Therefore, the notch should be place in different surfaces of the test sample to perform orientation dependent measurements. Nonetheless, previous results for other materials indicate that this is not a critical parameter as the effect of the building axis seems negligible for similar materials. This allowed us to manufacture all the tests samples with the same orientation, i.e. larger surface lying in the xy-plane, thus reducing the overall manufacturing time (as less layers are needed for fabrication). Consequently, all samples were machined with the notch at the top.

The results have been illustrated in section 7.3, each point represents the average energy needed to fracture 6 samples fabricated in 5 particular iterations, while red bars indicate standard deviation from this value.

The dashed line represents an average energy obtained from these 5 values. As it can be observed, the toughness of the test samples presents variations for the different iterations, but no pattern or trend can be appreciated. The values oscillate around 10 J (values not far to those observed in other high temperature resistance materials discusse by Kruth apparently randomly, and no correlation is either found with respect to the porosity. Actually, these variations are within the range of energies delimited by almost all the standard deviations observed so this is an expected behaviour.

7.5.3. Fatigue

Koutiri, I, et al [21, 18]investigated the influence of SLM process parameters on the fatigue behaviour of as-built Inconel 625 parts obtained by inert gas atomization, was supplied by PRAXAIR with a 10-45 μm particle range. This particle distribution was confirmed by a laser granulometer devicẹ (CILAS) as shown in figure 7.14. The particle shape has also been observed by Scanning Electron Microscopy (SEM). Particles present a mostly spherical shape with no satellites, and no agglomeration.

All the specimens investigated in this study are manufactured by a SLM 125HL machine provided by the SLM Solutions GmbH company. In the first part of the study, we will vary the irradiation parameters: laser power P (between 100 and 400 W), scanning velocity V (0.5to 1.8 m/s), beam diameter D (70 to 160 μm), for both the contour and hatching parameters (table 2). A specific focus will also be given to the influence of tilt angle θ (0° to 35°) between the platform and the part. The other process parameters (hatch distance, hatching strategy = stripes) will be fixed and will not be considered.

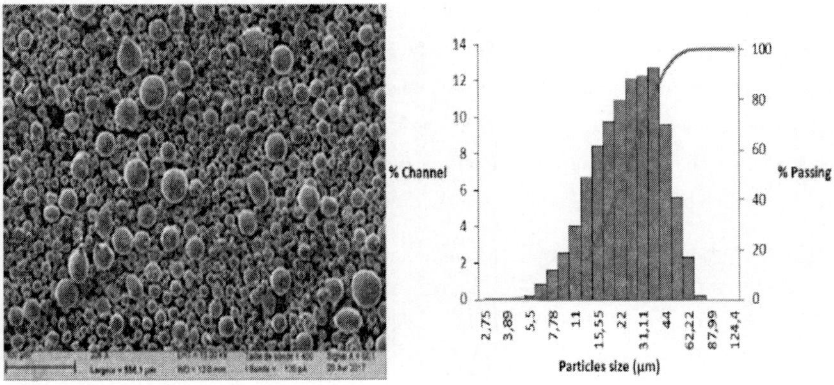

Figure 7.14: (a) SEM Analysis of Powder Particles, (b) Inconel 625 Grain Distribution [21, 18]

In this first stage, dedicated to the parametric study, only blocks of 10 x 18 x 4 mm were manufactured. For each set of parameters, a volume energy density (VED) has been calculated using equation 7.5. Such an equation allows considering the amount of energy induced by the laser scanning on the powder bed as a unified process parameter. This specific analytical formulation of laser energy input VED can be directly correlated to penetration depths of beads as shown analytically by Fabbro and confirmed experimentally in Gunenthiram.

$$\text{VED } (J/m^3) = P(W)/V\ (m/s) \times A(m^2) \qquad (7.2)$$

The second part of the study is dedicated to the production of tensile and fatigue samples, using optimized process parameters (P, V, D). The geometry of tensile and fatigue specimens are shown in Figure 7.15. A rounded edge was carried out to avoid local fatigue crack initiation.

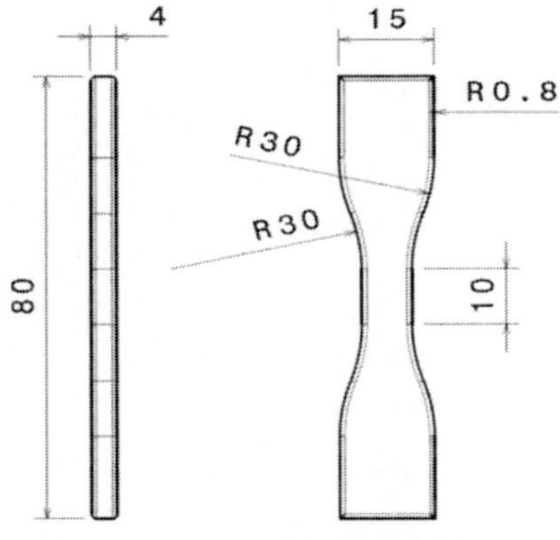

Figure 7.15: Geometry of the Specimens for Testing [21, 18]

SLM parameters were chosen for manufacturing fatigue samples, mostly considering the reduction of porosity rate and surface finish which are both expected to be major contributors to fatigue crack initiation. For this reason, SLM parameters, corresponding to high VED values (93 J/mm^3 for the contour and 87 J/mm^3 for the hatching), and a low building angle (10°) were used which is shown in table 7.3 base on the analysis of roughness with respect to process parameters and building angle which has a lowest value of roughness.

Table 7.3: SLM Process Parameters Carried Out on Static Tensile and Fatigue Samples [21,18]

θ°	Contour parameters			Hatching parameters		
	P (W)	V(m/s)	D(μm)	P (W)	V(m/s)	D(μm)
10	180	0.5	70	175	0.6	70

Seven fatigue test were conducted to determine the number of cycles to failure for as-built and polished specimens, at different stress amplitudes (with a 25 MPa increment). Even if the number of specimens is limited, the results and the failure surface obtained give important information and ways of improving SLM materials for fatigue applications. All tests were carried out at a test frequency of 20 Hz with a load R-ratio equal to -1 using the specimen geometry and the results are tabulated in table 7.4 and the corresponding curves in figure 7.16.

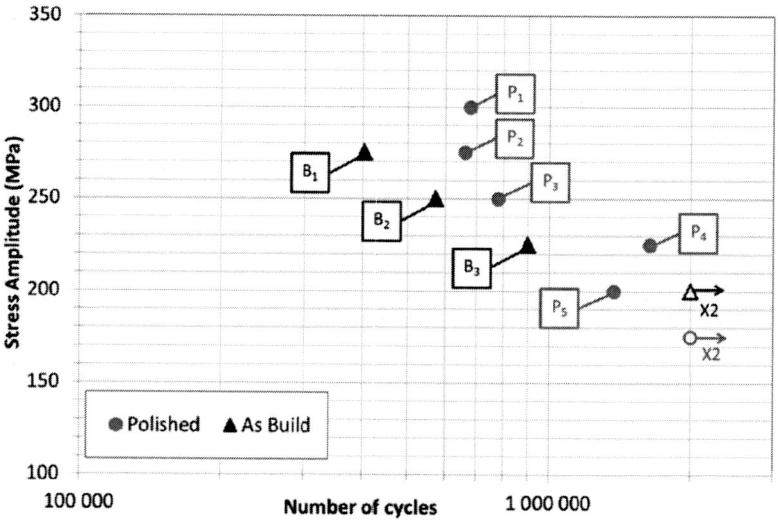

Figure 7.16: S-N Curves for Polished and as-built Batches [21,18]

Table 7.4: Fatigue Data, Fully Reversed Tensile Loads (R=-1) [21,18] p: Initiation from a Porosity, e:initiation from an Embedded Particle

Polished specimen			As- built specimen			
Specimen name	σ_a(MPa)	N(cycles)	Specimen name	σ_a(MPa)	N(cycles)	Sa(μm)
P1	300	678663	B1	275	403887[e]	8.56
P2	275	662916[p]	B2	250	573095[p]	8.56
P3	250	779395[p]	B3	225	901787[e]	10.69
P4	225	1641598	B4	200	2000000	5.46
P5	200	1373450[p]	B4	225	1069045[e]	
P6	175	2000000	B5	200	2000000	4.82
P6	200	1190074	B5	225	2000000	
P7	175	2000000	B5	250	355349	
P7	200	261270[p]				

Specimens that survived $2x10^6$ cycles have been retested at a higher stress level, increased by 25MPa, to estimate the fatigue strength and to observe the fatigue damage mechanism on the failure surfaces.

The estimated fatigue limits are between 175 and 200 MPa. This is much lower than the fatigue strengths obtained with other manufacturing processes. Considering previous "SLM and fatigue" studies on other classes of materials, like an AlSI10Mg with a fatigue limit equal to approximately 30% of the UTS for the same building orientation, the fatigue strengths determined here can also be considered to be very low.

The fatigue results clearly indicate an increase in the number of cycles to failure (i.e. an average of approximately $+2x10^5$ cycles for stress amplitudes of 225, 250 and 275 MPa) for the polished specimens, even if a higher difference could have been expected between the two conditions, due to the difference in surface roughness. However, for the same number of cycles, the difference in stress amplitude, observed between both conditions remains low. Despite the high surface roughness of as-built specimen, the difference in the fatigue strength between the polished and the as-built specimens, the difference in the fatigue strength between the polished and the as-built specimens is not pronounced (about 10%).

7.6. Tribological Behaviour of SLMed Parts

7.6.1. Wear

Jia, Q., & Gu, D, et al [19] made an investigation on Selective laser melting additive manufacturing of Inconel 718 super alloy parts: Densification, microstructure and properties along with tribological property with the gas atomized Inconel 718 powders with the spherical shape and the particle size in a range of 15–45 lm were used as the starting materials in this study. The chemical compositions of Inconel 718 powders are given in table 7.5.

Table 7.5: Chemical Compositions of Inconel 718 Powder (%wt.) [19]

Cr	Mo	Al	Ti	Fe	Nb	C	Ni
18.4	4.2	0.3	0.9	17.7	5.1	0.08	Balance

The SLM experimental setup consisted of an IPG Photonics Yitterbium YLR-200-SM fiber laser with a power of 200Wand a spot size of 70 μm, an automatic powder layering apparatus, an inert argon gas protection system and a computer system for process control. Based on a series of preliminary experiments, the laser power (P) was preset at 110, 110, 120, 130W and scan speeds (v) were set at 600, 400,400, 400 mm/s, respectively. Thus, four different linear laser energy densities η of180, 275, 300 and 330 J/m, which were defined by [20]:

$$\eta = P/v \tag{7.3}$$

Cubic specimens with dimensions of 8 mm × 8mm × 6mm were prepared using SLM as shown in figure 7.17.

Figure 7.17: Photograph of SLM-processed Inconel 718 Specimen [19]

Dry sliding wear tests were conducted in a HT-500 ball-on-disk tribometer at room temperature. A 3 mm diameter bearing steel GCr15 ball was taken as the counter face material, using a test load of 3 N. The friction unit was rotated at 560 rpm for 15 min and the rotation radius was 2 mm. The coefficient of friction (COF) was recorded during sliding. The wear volume (V) was determined gravimetrically using the relation given in equation 7.4.

$$V= M_{loss}/ \rho \qquad (7.4)$$

Where, M_{loss} was the weight loss of the specimens after wear tests and ρ the density of SLM-processed parts. The wear rate (ω) was calculated by equation 7.5:

$$\omega = V/WL \qquad (7.5)$$

Where, W was the contact load and L was the sliding distance.

The variations of COFs and wear rates of SLM-processed parts are provided in figure 7.18 and 7.19 respectively.

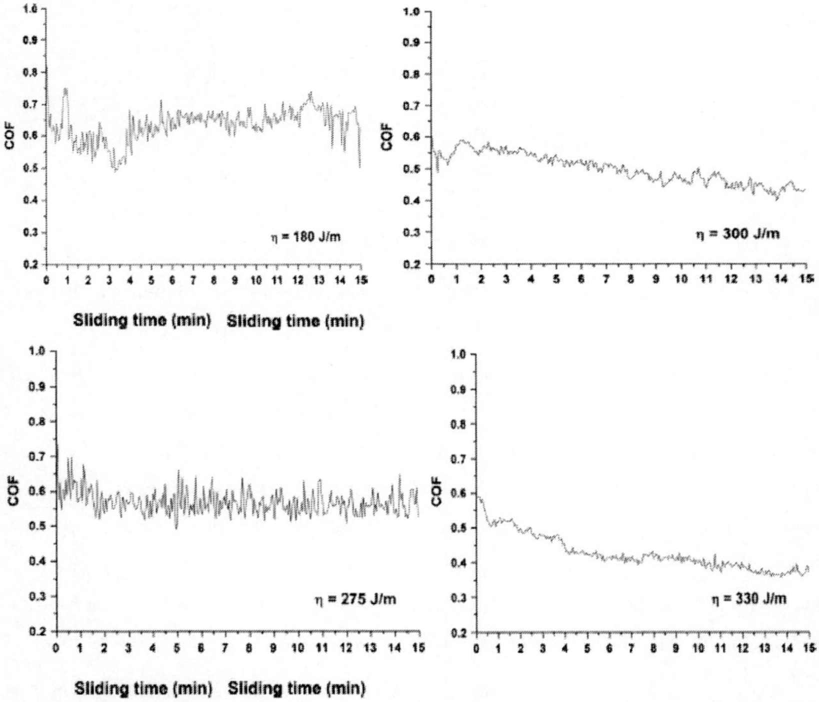

Figure 7.18: COFs of SLM-processed Inconel 718 Parts using various Processing Conditions [19]

It was apparent that the applied η exerted a key role in influencing the obtainable wear performance. On increasing g from 180 J/m to 330 J/m, the average COF decreased slightly from 0.62 to 0.58 and the attendant wear rate from 9.12×10^{-4} mm3/N m to 7.35×10^{-4} mm^3/N m. In addition, the COF values of the corresponding samples both showed apparent fluctuation behaviour with increasing sliding time. The average COF value decreased to 0.49 at a higher g of 300 J/m, leading to the reduced wear rate of 5.66 $\times 10^{-4}$ mm^3/N m. As the applied g increased to 330 J/m, the COF declined gradually until a steady state value of 0.36 was obtained, in which the wear rate reduced to the lowest of 4.64 $\times 10^{-4}$ mm^3/N m.

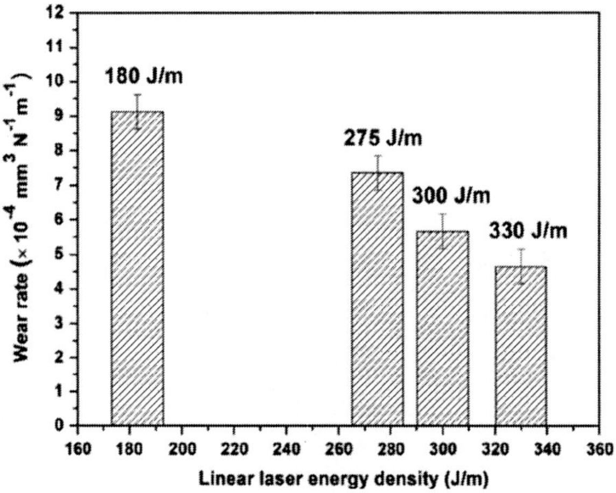

Figure 7.19: Wear Rates of SLM-processed Inconel 718 Parts using different Processing Parameters [19]

In general, the SLM processed Inconel 718 specimens exhibited relatively good wear performance without any grooves or cracks after friction. Nevertheless, as shown in figure 7.20, the worn surface morphologies varied greatly from the applied g. At a lower g of 180 J/m, the worn surface was considerably rough with evidence of observed granular wear debris and delamination seen in 7.20(a). Such microstructures indicated that the specimen experienced severe adhesive wear, which in turn resulted in a relatively higher wear rate as indicated in figure 7.19. In order to have a deep understanding of the frictional and wear characteristics of the corresponding parts, the morphologies of worn surfaces are illustrated in figure 7.20.

The worn surface became much smoother and free of any abrasive fragments as g increased to 275 J/m. In this circumstance, a mild adhesive wear was occurred at the contacts between the specimen and the GCr15 ball, causing the formation of adhesion tribolayer in local zone as indicated in figure 7.20(b). As g was elevated to 300 J/m, the consecutive adherent tribolayer was observed to be covered on the worn surface seen in figure 7.20(c). When an increased g of 330 J/m was settled properly, a smooth and dense wear track was formed, showing no significant fractures on the worn surface as depicted in figure 7.20(d).

During the wear tests, the counter face ball slid against the surface continuously, whereas the wear performance was primarily influenced by hardness and related microstructural features of corresponding parts [21]. At a lower laser energy density input, the limited hardness triggered the higher wear rate under the effect of mechanical shear force between the sliding surfaces. The marked fluctuation of COF values were attributed to the formation of coarsened columnar dendrites.

The wear resistance was enhanced owing to the refined microstructures as well as improved hardness, on increasing the applied laser energy densities. However, the clustered c dendrites gave rise to the fluctuations of COF values at this circumstance. The formed protective adherent tribolayer on worn surfaces, as shown in figure 7.20(c) and (d), made considerable contributions to the further improved wear performance of SLM-processed parts. Thus, it was reasonable to consider that as the input laser energy density increased, the dominant mechanism of material removal during sliding process probably changed from the sliding friction to the rolling friction, which produced the self-lubricating ability [22].

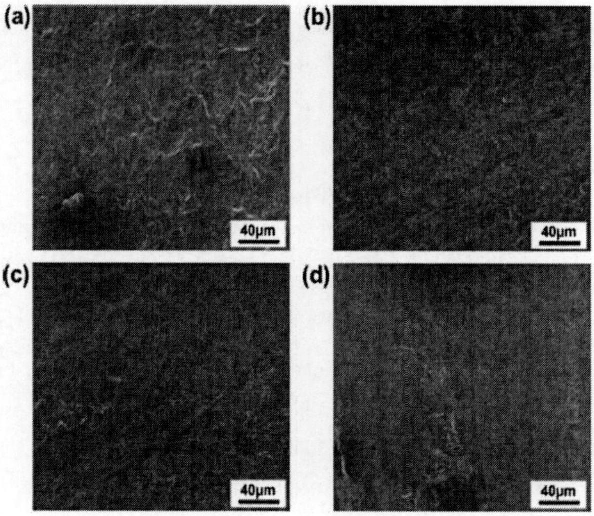

Figure 7.20: SEM Images Showing Morphologies of Worn Surfaces at various Laser Energy Densities: (a) 180 J/m; b) 275 J/m; (c) 300 J/m; (d) 330 J/m [19]

7.6.2. Corrosion

Luo, S et al [23] investigated the corrosion behaviours of Inconel 718 alloy produced by selective laser melting following different heat treatments along with microstructure evolution.

SLMed Inconel 718 specimens were produced by using self-developed LSNF-2 equipment. Based on the previous research, the optimal SLM process parameters (a laser power of 350 W, scanning speed of 800 mm/s, hatch spacing of 120 μm and layer thickness of 40 μm) were applied in this work. During SLM experiments, argon gas atomized spherical Inconel 718 alloy powders with an average size of about 32.7 μm were used as the original material. The chemical composition of powders measured using an inductively coupled plasma-atomic emission spectrometry (ICP-AES) are shown in table 7.6.

Table 7.6: Chemical Composition of Inconel 718 Powders after Process [23]

Element	Al	Ti	Cr	Mn	Fe	Mo	Nb	C	Ni
Wt.%	0.56	1.01	18.94	0.01	18.23	3.0	4.98	0.04	balance

To avoid the unfavourable effect of surface changes induced by SLM process and ensure the consistency of initial microstructures, cubic samples (10×10×3 mm) used for subsequent heat treatment and electrochemical tests were machined from the centre of the as-deposited bulk. The cubic specimens were annealed at 500 °C for 3 hours followed by furnace cooling to eliminate a part of residual stress [24]. In order to eliminate the influences of surface scratches on the corrosion resistance, the cubic specimens were polished. According to previous research [25], the heat treatment was performed in a vacuum furnace with the processes shown in Table 7.7.

For Inconel 718, dynamic recrystallization can occur due to the severe lattice distortion during hot rolling at 980 °C. As a result, the homogeneity of microstructures is improved and the grains are largely refined.

Table 7.7: Heat Treatment Process of Inconel 718 Produced by SLM [23]

No.	Processing
ST1	940°C × 1 hour Air cooling(AC)
ST2	980°C × 1 hour Air cooling(AC)
ST3	1020 °C × 1 hour Air cooling(AC)
SA(Solution plus double aging)	980 °C×1 h (AC) +720 °C×8 h (FC, 50 °C/h) +620 °C ×8 h (AC)

ST= Solution annealing treatment, SA = Solution plus double aging, FC= furnace cooling, AC=Air cooling

That is to say, the hot-rolled Inconel 718 exhibits excellent corrosion resistance. Thus, a rolling Inconel 718 sample hot-rolled at 980 °C was selected to make a comparison.

The electrochemical tests were conducted at room temperature using a CorrTest CST310 electrochemical station. A traditional three-electrode cell containing a KCl-saturated calomel electrode, a platinum counter electrode, as well as the cubic specimen serving as a working electrode with a surface of $100mm^2$ in cross-section was applied in the electrochemical measurements. The polarization experiments and electrochemical impedance spectroscopy (EIS) were performed after open circuit potential (OCP) is stabilized by immersing the electrode for enough time in 3.5 wt % NaCl solution. The corresponding parameter variables are shown in table 7.8 and the graphs were shown in figure 7.21.

To guarantee the data reproducibility, each electrochemical test was repeated at least three times on different samples with the same exposed surface areas and initial microstructures. CorrTest software involving ZView is equipped on the electrochemical workstation, which can be used to determine the EIS curve fitting.

Table 7.8: Electrochemical Test Parameter Variables [23]

Test	Parameter	Value
Dynamic polarization	Sweep range(V vs. OCP)	-0.5 to 1.51
	Sweep rate(mV/s)	1
EIS	Frequency range(Hz)	10^{-2} to 10^{5}
	Signal	Sinusoidal
	Amplitude(mV)	10

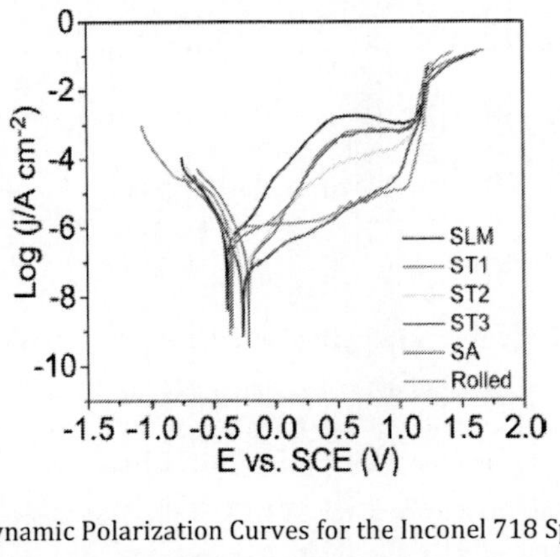

Figure 7.21: Potentiodynamic Polarization Curves for the Inconel 718 Specimens Fabricated by different Processes in 3.5 wt% NaCl Solution [23]

Figure 7.21 depicts the comparisons of Potentiodynamic polarization curves for the SLM, SLM following various heat treatments, as well as rolled samples in 3.5 wt% NaCl solutions. Figure 7.22 summarizes the average electrochemical parameters: (a) corrosion potential and corrosion current density; (b) corrosion rate; (c) passive potential and passive current density and (d) pitting potential. Derived from the Tafel slope method and corresponding standard deviations. Obviously, the SLM sample possesses the lowest corrosion potential (~−0.45 V), and the corrosion potential increases in an order of SLM<ST1<rolled<ST2<ST3<SA.

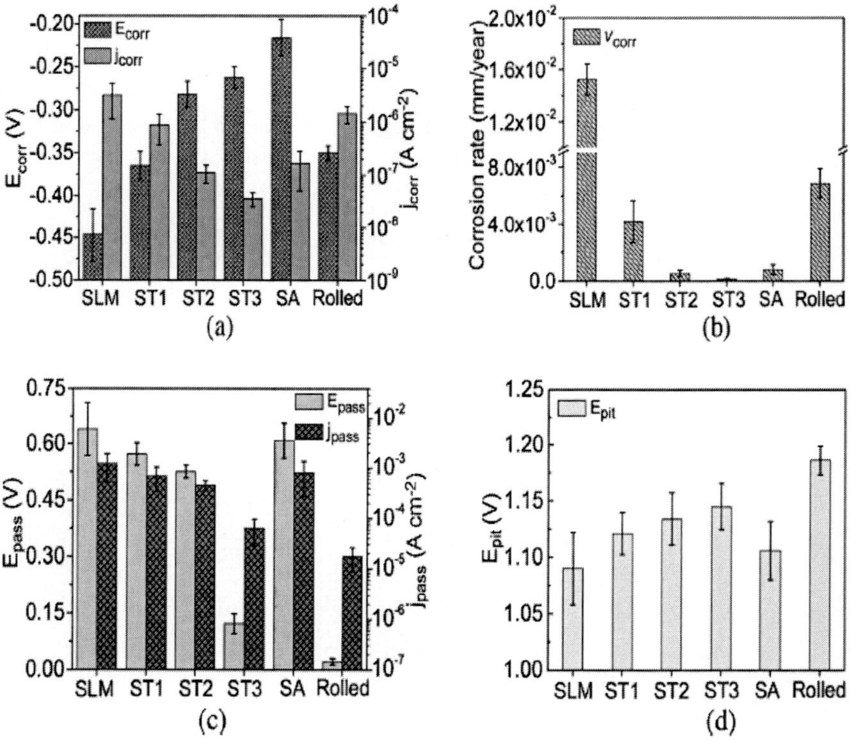

Figure 7.22: Inconel 718 Specimens Fabricated by different Processes in 3.5 wt% NaCl Solution with various Potential [23]

However, corrosion current densities for the SLM, ST1 and rolled samples are nearly one order of magnitude higher than those of the ST2, ST3 and SA ones. Meanwhile, the ST3 sample produces the lowest corrosion current density (only 3.61×10−8 A cm^{-2}), while that of the SLM samples is the largest (as high as 3.18×10−6 A cm^{-2}). Based on the corrosion current density, the corrosion rate can be calculated by using the Faraday law [26,]:

$$v_{corr}\,(mm/year) = 0.00327 \times EW \times j_{corr} / D \qquad (7.6)$$

Where, EW is the equivalent weight of Inconel 718 alloy, and D is the density (8.24 g/cm^3). According to equation 7.6, there is a positive correlation between corrosion rate and corrosion current density. Thus, the corrosion rate decreases in an order of SLM>rolled>ST1> SA>ST2>ST3, as shown in figure 7.22(b). As can be seen from the polarization curves, the ST3 and rolled samples exhibit a tendency of being spontaneously passive. While the SLM, ST1, ST2 as well as SA samples all exhibit a typical characteristic of active dissolution, and the steady-state passivation regions are formed until the potential continuously increases to a large value. Also, the passive potential decreases in an order of SLM>SA>ST1> ST2>ST3>rolled, as depicted in figure 7.22(c).

Meanwhile, the passive current densities of the SLM, SA, ST1 and ST2 samples are significantly higher than those of the ST3 and rolled ones. Similarly, the rolled sample possesses the smallest passive current density (only 1.74×10−5 A cm^{-2}). These illustrate that the surface passive film is much easier to form on the surface to inhibit the further corrosion for the rolled sample. Epit represents the breakdown potential where the current suddenly increases, which is usually applied to estimate the resistance to pitting corrosion. Thus, the larger the value of Epit, the more stables the surface passive film becomes. From figure 9d, it is observed that the breakdown potential exhibits in an order of rolled> ST3>ST2>ST1>SA>SLM, which reveals that the surface passive films of rolled and ST3 samples present the best pitting resistance against the attack of Cl− ion. Therefore, the corrosion resistance of the Inconel 718 specimens obtained by different processes in 3.5 wt% NaCl solution follows in the order of rolled>ST3>ST2>ST1>SA>SLM based on the above results.

It has been concluded that the ST3 and rolled samples possess better corrosion resistance than other samples. To further compare the corrosion behaviors of ST3 and rolled samples, the general corrosion resistances of them exposed to 3.5 wt% NaCl solution were measured by using EIS after OCPs are stabilized.

The typical EIS measurements for the ST3 and rolled Inconel 718 samples in 3.5 wt% NaCl solution: (a) best fitting Nyquist plots; (b) best fitting Bode plots; (c) equivalent circuit used for impedance spectra analysis displayed by Nyquist and Bode plots are shown in figure 7.23(a) and (b), respectively.

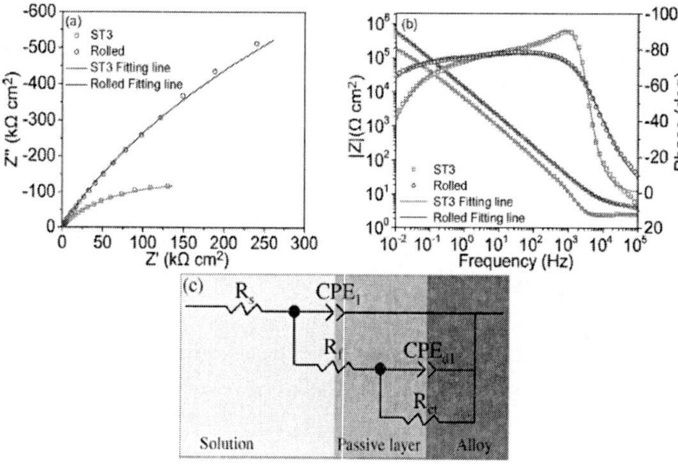

Figure 7.23: Typical EIS Measurements for the ST3 and Rolled Inconel 718 Samples in 3.5 wt% NaCl Solution [23]

The fitting lines are in close proximity to the original data with a mean error lower than 1%. Apparently, the Nyquist plot diameter of the rolled sample is larger than that of the ST3 sample, meaning a stronger corrosion resistance and lower activity. Usually, a duplex structure oxide film containing a compact inner layer and a porous external layer forms on the surface for passivated metals [27, 28] occurs. In addition, as can be observed in figure 7.23(b), plateaus appear in the Bode plots over wide ranges of frequency. Thus, an electrical equivalent circuit (EEC) with two time-constants was applied to model the EIS data, as depicted in figure 10c. The EEC is composed of solution resistance (R_s), film resistance (R_f) and charge transfer resistance (R_{ct}), along with constant phase elements (CPEdl and CPE1). Note that the CPEdl was used owing to the imperfect flat of the electrode surface [29]. This kind of EEC has been largely applied in recent studied alloys [30].

Table 7.9: EIS Measurement Readings [23]

Samples	**R_f (kΩcm²)**	**CPE_1 (μΩ⁻¹cm⁻² s⁻ⁿ)**	**n_1**	**R_{ct} (kΩcm²)**	**CPE_{dl} (μΩ⁻¹cm⁻² s⁻ⁿ)**	**n_2**
ST3	0.05	1.39	0.9970	333.16	33.79	0.7901
Rolled	90.96	13.91	0.8740	5234.90	2.63	0.5887

Table 7.9 summarizes the fitted results of EIS measurements using nonlinear least square (NLS). Because the value of R_f just reveals the resistance for the charge transfer in the diffusion layer (a relatively porous outer layer), while the value of R_{ct} can represent the resistance for the

charge transfer in the double layer (passive film), the simulated values of R_{ct} are markedly higher than those of R_f for both samples. Moreover, the values of R_f and R_{ct} for the rolled sample are larger than those of the ST3 sample. Specially, the higher R_{ct} implies that the corrosion dissolution of surface passive film is more difficult Therefore, the corrosion resistance of the rolled sample is superior to ST3 sample, which accords with the result acquired from the polarization analysis. Thus, it can be concluded that the corrosion resistance of SLMed sample will have significant improvement after solution annealing treatment, but is still lower than that of rolled one. Further optimization of heat treatment process is needed to make the corrosion resistance of SLMed sample comparable to rolled one.

7.7. Defects in SLM of Ni based Alloys

7.7.1. Micro Crack and Porosity

Maciej Mazur et al [31] and his team investigated the Selective laser melting of Inconel 625 alloy with reduced defect formation with the use of gas atomized powder with different size 14.7, 33, 57.1, 25.2 and 34.7 μm with respect to % volume density. SLM machine with 400 W fibre lasers was used. The processing parameters of Inconel 625 specimens has a hatch distance of 0.12mm and layer thickness of 0.03mm for all samples with and without skywriting and check board scan pattern as shown in figure 7.24 for 200W 150W, 250W power, 900mm/s, 690mm/s, with energy densities are done as shown in table 7.10.

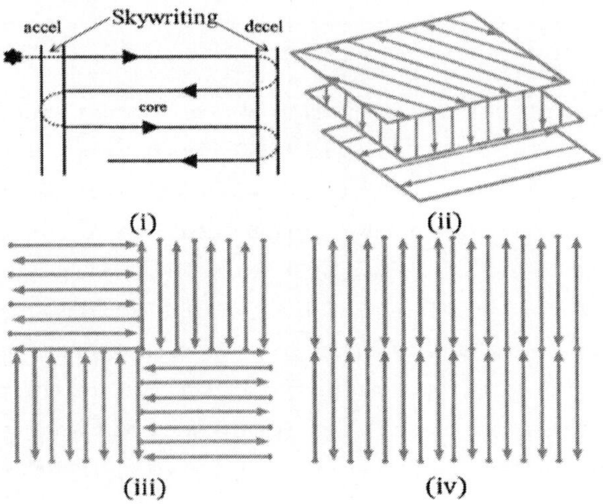

Figure 7.24: Laser Scan Strategies: (i) Skywriting Function (Ref. 10), (ii) Layer Rotation by 66.7°, (iii) Checkerboard Pattern, and (iv) Stripe Pattern [31]

Table 7.10: SLM Process Parameters for Inconel 625 Specimens [31]

Process cond.ID	Power (W)	Scan speed (mm/s)	Energy density (J/mm³)	Tensile test	Special conditions
P1SRC	200	900	61.7	N	Skywriting disabled, layer rotation 0°, checkerboard scan pattern
P1RC	200	900	61.7	N	Skywriting enabled, layer rotation 0°, checkerboard scan pattern
P1SR	200	900	61.7	Y	Skywriting disabled, layer rotation 0°, stripe scan pattern
P1S	200	900	61.7	Y	Skywriting disabled, layer rotation 66.7°, stripe scan pattern
P1R	200	900	61.7	Y	Skywriting enabled, layer rotation 0°, stripe scan pattern
P1	200	900	61.7	Y	-
P2	150	690	60.4	N	-
P3	250	1150	60.4	N	-
P5	150	520	80.1	N	-
P4	200	690	80.5	N	Skywriting enabled, layer rotation 66.67°, stripe scan pattern
P6	250	865	80.3	N	-
P8	150	415	100.4	N	-
P7	200	555	100.1	N	-
P9	250	690	100.6	N	-

As an initial trial, prismatic test specimens were manufactured with the parameters corresponding to P1SRC in table 7.10. Figure 7.25(i) shows the corresponding micrographs, where porosity defects are present at locations corresponding, predominantly to the laser scan track start and end points aligned with the applied checkerboard scan strategy. Subsequent samples manufactured with the skywriting function enabled (P1RC in Table 7.10) did not exhibit such porosity defects indicated in figure 7.25(ii). As such, the defects are most likely attributable to a non-uniform laser exposure introduced in the absence of the skywriting function indicated in figure 7.24(ii). Such non uniformity is typically introduced by the acceleration and deceleration of the scanning laser mirror galvanometer at the start and ends of scan tracks, which can thereby increase volumetric energy density levels compared with the central track sections.

The increased energy levels can result in fusion defects and increased porosity. The skywriting function compensates for the inertia of the laser mirror system by coordinating the laser on and off operation throughout a scan track to only enable the laser during periods of constant mirror velocity.

The presence of porosity defects associated with the scan strategy was confirmed with micro-CT scans of manufactured tensile test specimens.

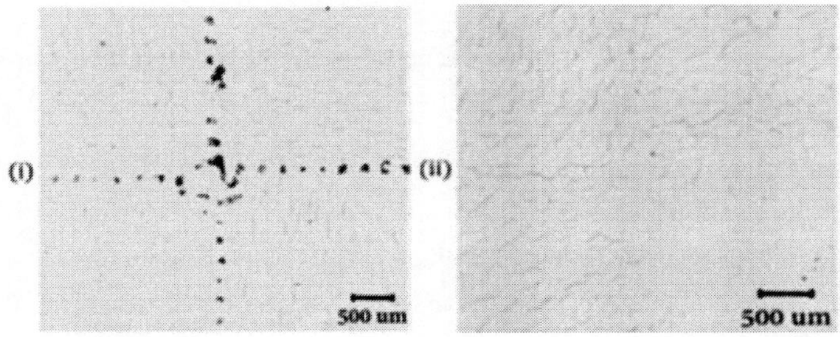

Figure 7.25: (i) P1SRC Showing Porosity and Micro Crack Defects and (ii) P1RC without Periodic Porosity Defects for Prismatic Speciemen [31]

Figure 7.26: Micro-CT Images for Tensile Samples Gauge Sections Indicating Porosity Defects over 50 μm in Size [31]

Figure 7.26 shows scans of 6 mm gauge sections of tensile samples indicating porosity defects larger than approximately 50 μm in size (i) no skywriting with no layer rotation (P1SR), (ii) no skywriting with layer rotation (P1S), and (iii) skywriting with layer rotation (P1). Periodic defects are visible for process condition P1SR (skywriting off, layer rotation off) distributed defects for condition P1S (skywriting off, layer rotation on), and the absence of defects (within the measurement resolution of the CT scanner) in process condition P1 (skywriting on, layer rotation on).

The absence of the skywriting function and layer rotation resulted in defects accumulating in periodic planes with 1mm intervals. This corresponds to a standard scan pattern layer offset of 1mm used in a stripe pattern scan strategy employed by the SLM Solution 125 system used to manufacture the test samples. As such, it appears that IN625 alloy is sensitive to the energy density and that use of the skywriting function is recommended. In addition to porosity defects, micro cracks are observed in figure 7.25 which appear to be aligned with the scan orientation within each quadrant of the applied checkerboard scan strategy.

Microstructures from the XY plane (i.e. perpendicular to the build direction) can be seen in figure 7.27 for prismatic specimens with processing parameters corresponding to P1–P9.

Fine cracks can be seen in the microstructure for each condition, and in several cases, the cracks appear to be fairly aligned. Preliminary measurements from the micrographs indicate that the average area coverage of the cracks ranges between 0.38% and 1.06% for all process parameter combinations studied. Similarly, an increase in porosity can be seen as the energy density increased from 80 to 100 J/mm^3, particularly when the laser power was 150 or 200W.

The pores are 50 μm in diameter or less and are mostly spherical. The tensile test results averaged from three samples per process condition were discussed and concluded that the representative stress–strain response results indicate that the presence of porosity introduced by the suboptimal scan strategy has a significant impact on all mechanical properties, particularly the elongation. However, when comparing the mechanical properties to reference values for IN625, the presence of micro cracking does not appear to have a significant effect on the quasistatic tensile mechanical properties.

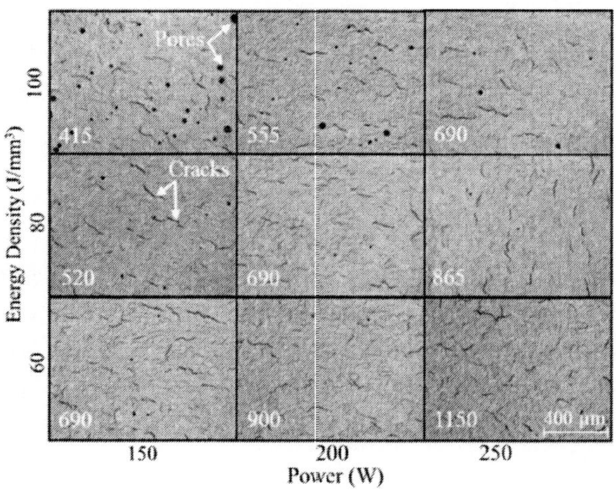

Figure 7.27: Microstructure in the XY Plane as a Function of Laser Power and Energy Density [31]

References

[1] https://www.langleyalloys.com/en/what-is-inconel/

[2] https://www.corrotherm.co.uk/blog/5-applications-of-inconel- alloys.

[3] Choi, Joon-Phil; Shin, Gi-Hun; Yang, Sangsun; Yang, Dong-Yeol; Lee, Jai-Sung; Brochu, Mathieu; Yu, Ji-Hun, "Densification and microstructural investigation of Inconel 718 parts fabricated by selective laser melting", *Powder Technology,* 310(), 2017, pg. 60–66.

[4] ASTM B213-03, *"Standard test method for flow rate of metal powders",* American Society for Testing and Materials, West Conshohocken, Pennsylvania, 2003.

[5] ASTM B212-09, *"Standard test method for apparent density of free-flowing metal powders using the Hall flow meter funnel",* American Society for Testing and Materials, West Conshohocken, Pennsylvania, 2009.

[6] D.D. Gu, W. Meiners, K. Wissenbach, R. Poprawe, "Laser additive manufacturing of metallic components: materials, processes and mechanisms", *Int. Mater. Rev.,* 57, 2012, pg. 133-164.

[7] Y. Li, D.D. Gu, "Parametric analysis of thermal behavior during selective laser melting additive manufacturing of aluminum alloy powder", *Mater. Des,* 63, 2014, pg. 856-867.

[8] J.P. Kruth, G.N. Levy, F. Klocke, T.H.C. Childs, "Consolidation phenomena in laser and powder-bed based layered manufacturing", *CIRP Ann. Manuf. Technol.,* 56, 2007, pg. 730-759.

[9] N.B. Dahotre, S. Harimkar, *Laser fabrication and machining of materials,* Springer, New York, 2008.

[10] W.E. King, H.D. Barth, V.M. Castillo, G.F. Gallegos, J.W. Gibbs, D.E. Hahn, C. Kamath, A.M. Rubenchik, "Observation of keyhole-mode laser melting in laser powder-bed fusion additive manufacturing", *J. Mater. Process. Technol.* 214, 2014, pg. 2915-2925.

[11] Ardila, L.C., Garciandia, F., González-Díaz, J.B., Álvarez, P., Echeverria, A., Petite, M.M., Ochoa, J, "Effect of IN718 Recycled Powder Reuse on Properties of Parts Manufactured by Means of Selective Laser Melting", *Physics Procedia,* 56, 2014, pg. 99–107.

[12] J.A. Cherry, H.M. Davies, S. Mehmood, N.P. Lavery, S.G.R. Brown, J. Sienz, "Investigation into the effect of process parameters on microstructural and physical properties of 316L stainless steel parts by selective laser melting", *Int. Adv. Manuf. Technol.,* 76, 2015, pg. 869-879.

[13] E.O. Olakanmi, "Selective laser sintering/melting (SLS/SLM) of pure Al, Al–Mg, and Al–Si powders: Effect of processing conditions and powder properties", *J. Mater. Process. Technol.,* 213, 2013, pg. 1387-1405.

[14] J.R. Davis, "ASM specialty handbook: heat-resistant materials", *ASM International,* Materials Park, Ohio, 1997.

[15] J.R. Davis, "ASM specialty handbook: nickel", cobalt, and their alloys', *ASM International,* Materials Park, Ohio, 2000.

[16] B. Bazaz, A. Zarei-Hanzaki, S.M. Fatemi-Varzaneh, "Hardness and microstructure homogeneity of pure copper processed by accumulative back extrusion", *Mater. Sci. Eng. A,* 559, 2013, pg. 595-600.

[17] Ardila, L.C., Garciandia, F., González-Díaz, J.B., Álvarez, P., Echeverria, A., Petite, M.M., Ochoa, J, "Effect of IN718 Recycled Powder Reuse on Properties of Parts Manufactured by Means of Selective Laser Melting", *Physics Procedia, 56,* 2014, pg. 99–107.

[18] Koutiri, I., Pessard, E., Peyre, P., Amlou, O., & De Terris, T, "Influence of SLM process parameters on the surface finish, porosity rate and fatigue behavior of as-built Inconel 625 parts". *Journal of Materials Processing Technology,* 255, 2018, pg. 536–546.

[19] Jia, Q., & Gu, D, "Selective laser melting additive manufacturing of Inconel 718 superalloy parts: Densification, microstructure and properties", *Journal of Alloys and Compounds,* 585, 2014, pg. 713–721.

[20] Gu, D., & Shen, Y, "Effects of processing parameters on consolidation and microstructure of W–Cu components by DMLS", *Journal of Alloys and Compounds,* 473(1-2), 2009, pg. 107–115.

[21] Sheng, L.Y., Yang, F., Xi, T.F., & Guo, J.T, "Investigation on microstructure and wear behaviour of the NiAl–TiC–Al_2O_3 composite fabricated by self-propagation high-temperature synthesis with extrusion", *Journal of Alloys and Compounds,* 554, 2013, pg. 182–188.

[22] Niu, M., Bi, Q., Zhu, S., Yang, J., & Liu, W, "Microstructure, phase transition and tribological performances of Ni3Si-based self-lubricating composite coatings", *Journal of Alloys and Compounds,* 555, 2013, pg. 367–374.

[23] Luo, S., Huang, W., Yang, H., Yang, J., Wang, Z., & Zeng, X, "Microstructural evolution and corrosion behaviours of Inconel 718 alloy produced by selective laser melting following different heat treatments", *Additive Manufacturing,* 30, 2019, 100875.

[24] P.F. Guo, X. Lin, J.Q. Li, Y.F. Zhang, M.H. Song, W.D. Huang, "Electrochemical behavior of Inconel 718 fabricated by laser solid forming on different sections", *Corros. Sci.,* 132, 2018, pg. 79–89.

[25] W.P. Huang, J.J. Yang, H.H. Yang, G.Y. Jing, Z.M. Wang, X.Y. Zeng, "Heat treatment of Inconel 718 produced by selective laser melting: microstructure and mechanical properties", *Mater. Sci. Eng. A* 750, 2019, pg. 98–107.

[26] J.J. Yang, H.H. Yang, H.C. Yu, Z.M. Wang, X.Y. Zeng, "Corrosion behavior of additive manufactured Ti-6Al-4V alloy in NaCl solution", *Metall. Mater. Trans. A* 48, 2017, pg. 3583–3593.

[27] B. Wu, Z. Pan, S. Li, D. Cuiuri, D. Ding, H. Li, The anisotropic corrosion behaviour of wire arc additive manufactured Ti-6Al-4V alloy in 3.5% NaCl solution, *Corros. Sci.* 137, 2018, pg. 176–183.

[28] K. Kuwabara, K. Yamanaka, H. Shiratori, Y. Koizumi, A. Chiba, T. Fujieda, Mechanical and corrosion properties of AlCoCrFeNi high-entropy alloy fabricated with selective electron beam melting, *Addit. Manuf.* 23 (2018) 264–271.

[29] N. Dai, L. Zhang, J. Zhang, Q. Chen, M. Wu, "Corrosion behavior of selective laser melted Ti-6Al-4V alloy in NaCl solution", *Corros. Sci,* 102, 2016, pg. 484–489.

[30] A.H. Ettefagh, S. Guo, Electrochemical behavior of AISI316L stainless steel parts produced by laser-based powder bed fusion process and the effect of post annealing process, *Addit. Manuf.* 22, 2018, pg. 153–156.

[31] Maciej Mazur, Michael Benoit, Mark Easton, and Milan Brandt, "Selective laser melting of Inconel 625 alloy with reduced defect formation", *Journal of Laser Applications,* 32, Issue 2, 2020, 022058 (pg. 1-6).

Chapter VIII

Selective Laser Melting (SLM) of Stainless Steel

Chapter VIII

8. Selective Laser Melting (SLM) of Stainless Steel

8.1. Introduction

Stainless steels are iron-base alloys that contain a minimum of about 12% Cr, the amount needed to prevent the formation of rust in unpolluted atmospheres (hence the designation stainless). Few stainless steels contain more than 30% Cr or less than 50% iron. They achieve their stainless characteristics through the formation of an invisible and adherent chromium-rich oxide film. This oxide forms and heals itself in the presence of oxygen. Other elements added to improve particular characteristics include nickel, manganese, molybdenum, copper, titanium, silicon, niobium, aluminium, sulphur, and selenium. Carbon is normally present in amounts ranging from less than 0.03% to over 1.0% in certain grades.

Stainless steels fall into several general classes: austenitic, ferritic, martensitic, duplex, and precipitation hardened. The distinction between each class is based primarily on the predominant phase present in the stainless steel as determined by the major alloying elements.

The major alloying elements in stainless steels are chromium and nickel. Chromium primarily provides corrosion resistance and additional strength. Nickel provides strength and some corrosion resistance. Minor alloying elements include manganese, carbon, and molybdenum. Manganese is present in steels in small quantities, but at higher concentrations it stabilizes austenite and partially replaces nickel in the 200-series steels. Carbon is largely an impurity in austenitic steel, but it is a strengthening element in ferritic and martensitic steel, much as it is for carbon and low-alloy steel. Molybdenum provides additional strength and resistance to chloride pitting. Other elements, such as titanium or niobium, serve other purposes specific to the application for which the alloy was developed.

400-series steels were the first versions of stainless steel. They include the ferritic and martensitic grades that contain only chromium as a major alloying element, making them less expensive than austenitic grades. They are magnetic, generally more resistant to chloride attack than 300-series alloys, and some grades may be strengthened by heat treating.

Type 410 contains about 12% chromium. The ability to strengthen this alloy by heat treating to form martensite makes it a martensitic grade. Its low chromium content provides modest corrosion resistance. Given enough time, exposure to weather will cause it to rust. Type 430 is a ferritic grade that contains about 17% chromium. It cannot be strengthened by heat treating. 400-series stainless steels are generally more resistant to chloride attack than 300-series.

With sufficient quantities of nickel, stainless steel remains austenite at room temperature, creating the austenitic steels. They are nonmagnetic and cannot be heat treated for through hardening like carbon steels because the phase transformation to martensite does not occur in these alloys. The primary reason for their use is their superior resistance to corrosion in the atmosphere and aggressive chemical environments compared to 400-series.

300-series alloys contain chromium and nickel, and are the most popular austenitic grades. Types 301 and 304 are the most common alloys in use and are for general use. They contain 18% chromium, 9% to 10% nickel, and up to 0.15% carbon (301) or 0.08% carbon (304) as an impurity. Other 300-series alloys are modified versions of these alloys to achieve specific properties. Type 316 contains 2%-3% molybdenum to improve the resistance to corrosion in chloride-containing environments. Types 304L, 316L and other L-grades contain reduced carbon, less than 0.03%, to avoid microstructure changes during welding and other thermal processes which can damage the corrosion resistance. This detrimental change is known as sensitization. Types 321 and 347 contain small amounts of titanium and niobium, respectively, to prevent sensitization. They are capable of service at elevated temperatures, while the L-grades are intended to resist sensitization during fabrication. The image below shows the austenite grains in a 304 alloy.

Duplex stainless steels allow savings in material costs in corrosive applications such as chemical processing, including chloride- and sulphur-bearing environments. They consist of a mixture of austenite and ferrite in roughly equal proportions. Duplex stainless steels are subdivided into lean, standard, super, or hyper duplex based on the quantity of alloying elements. Duplex stainless steels contain more chromium and less nickel than 300-series and typically include nitrogen as an additional austenite stabilizer and molybdenum for corrosion resistance. 2205 (22% chromium, 5% nickel, and 3% molybdenum) is a common standard duplex stainless steel, and 2507 (25% Cr, 7% Ni plus 4% Mo) is a common super-duplex steel. The micrograph below shows duplex stainless steel[1].

The broad array of stainless steels available provides a vast portfolio of capabilities. However, each alloy has distinct advantages and disadvantages. When selecting a grade of stainless steels available it is important to consider how components will be fabricated and joined together, the specific environment to which it will be exposed, in addition to the considerations common to other alloys such as mechanical requirements and cost.

Stainless steels are used in a wide variety of applications. Most of the structural applications occur in the chemical and power engineering industries, which account for more than a third of the market for stainless steel products. These applications include an extremely diversified range of uses, including nuclear reactor vessels, heat exchangers, oil industry tubulars, components for chemical processing and pulp and paper industries, furnace parts, and boilers used in fossil fuel electric power plants. The relative importance of the major fields of application for stainless steel products are as shown as follows [2]:

Table 8.1: Usage of Stainless Steel [2]

Application	Percentage
Industrial equipment	
Chemical and power engineering	34
Food and beverage industry	18
Transportation	9
Architecture	5
Consumer goods	
Domestic appliances, household utensils	28
Small electrical and electronic appliances	6

8.2. Densification Behaviour based on Laser Related Process Factors

Li, R., et al [3] along with his teammates investigated the densification behaviour of gas and water atomized 316L stainless steel powder during selective laser melting. He analysed gas atomized powder possess spherical shape with the particle size distribution of 3-4μm and mean particle size of 20μm. Water atomized powder possess irregular shape with the particle size distribution of 6-50 μm with mean particle size of 30μm which is depicted in figure 8.1 with the help of SEM images of SS316L powders.

The process is carried using process parameters, where laser power (P) of 60–200 W, scan speed (v) of 20–300 mm/s, hatch spacing (h) of 0.1–0.8mm and layer thickness (d) of 0.05–0.1mm were used.

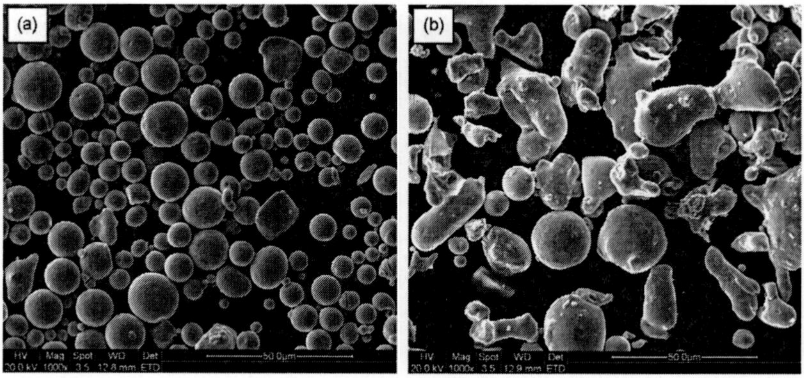

Figure 8.1: SEM Image of SS316L (a) Gas Atomized Powder Regular (b) Water Atomized Irregular Shaped Powder Particles [3]

It can also be found that this molten track was present in a high wetting angle, showing worse wetting ability. With the increase of laser power, the flow ability became better and the corresponding wetting angle was reduced, showing much better melting characteristic which was favourable for SLM process. In addition, an enhanced laser power can also cause a more deep penetration depth. The different wetting features of molten pools under varied laser powers could be understood by considering the correction between laser power and laser-induced liquid viscosity. The worse wetting characteristic was highly related to a high liquid viscosity, which was directly governed by the laser power input. A lower laser energy input could usually induce a higher capillary molten track. Enhance the laser power could enable a higher laser energy input, resulting in a higher melting temperature, which directly induce a much lower viscosity and resultant big-sized molten pool, favourable wetting, spreading and flattening characteristics is observed using (a) P=60W; (b) P=80W; (c) P = 100 W. v = 50mm/s is constant process parameters.

Figure 8.2 shows the characteristic microstructures of longitudinal sections of single scan tracks prepared by different scan speeds. At a scan speed of 150 mm/s, the scan track section showed a hump shape shown in figure 8.2(a), combined with a powder particle cohered in melting and solidification process as depicted in figure 8.2(b). It was caused by the insufficient liquid spreading ability on metal substrate. At a relative lower scan speed of 60 mm/s, the size of scan track section was obviously increased is clearly shown in figure 8.2(c). With the further decrease of scan speed to 20 mm/s, the molten pool tended to spread on the substrate, coupled with an increased melting width and depth as shown in figure 8.2(d). The different melting conditions of molten pools under varied scan speeds were also attributed to the different energy input.

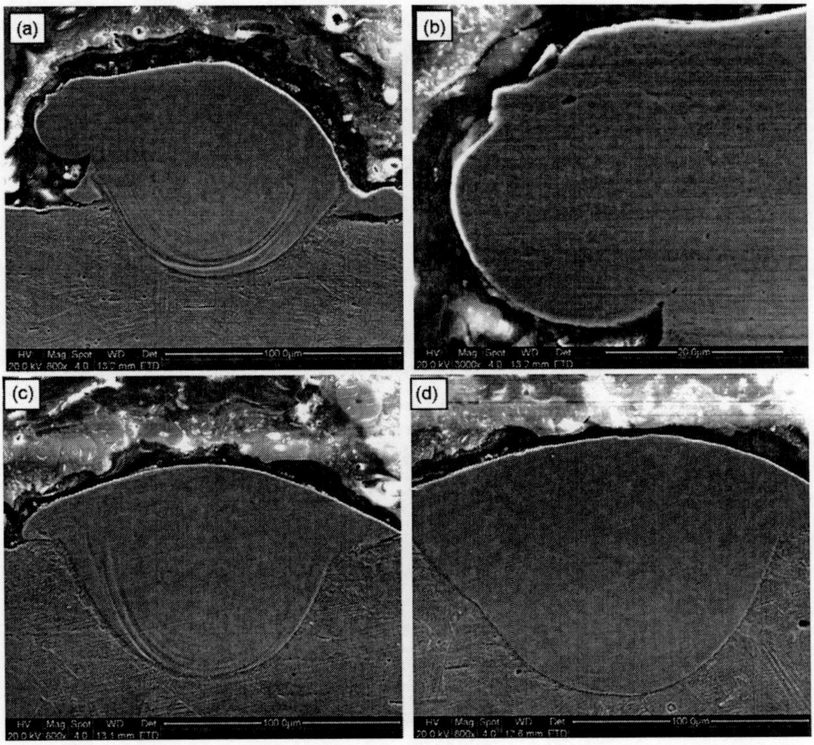

Figure 8.2: SEM Image at (a) v = 150 mm/s; (b) is Local Magnification of (a). (c) v = 60 mm/s; (d) v = 20 mm/ s P = 100W Constant [3]

Figure 8.3 explains that the surface morphologies of SLM samples prepared from the gas atomized powder using varied processing parameters, are with respect to figure 8.3 as follows (a) P=98W, d = 0.05mm; (b) is local magnification of (a); (c) P=80W, d = 0.05mm. v = 90mm/s and h = 0.1mm is constant. It can be seen that the processing parameters have a pronounced effect on the pores and agglomerates features. Figure 8.3(a) showed a relative smooth surface with very little visible pores, which was produced under a high laser power. Moreover, it exhibited that the well-arranged laser scan tracks were parallel to the scan direction, between which coherent bonding characteristics were obtainable. In more detail, a high-magnification SEM image shows that a large amount of small metallic balls were present on the sample surface shown in figure 8.3(b).

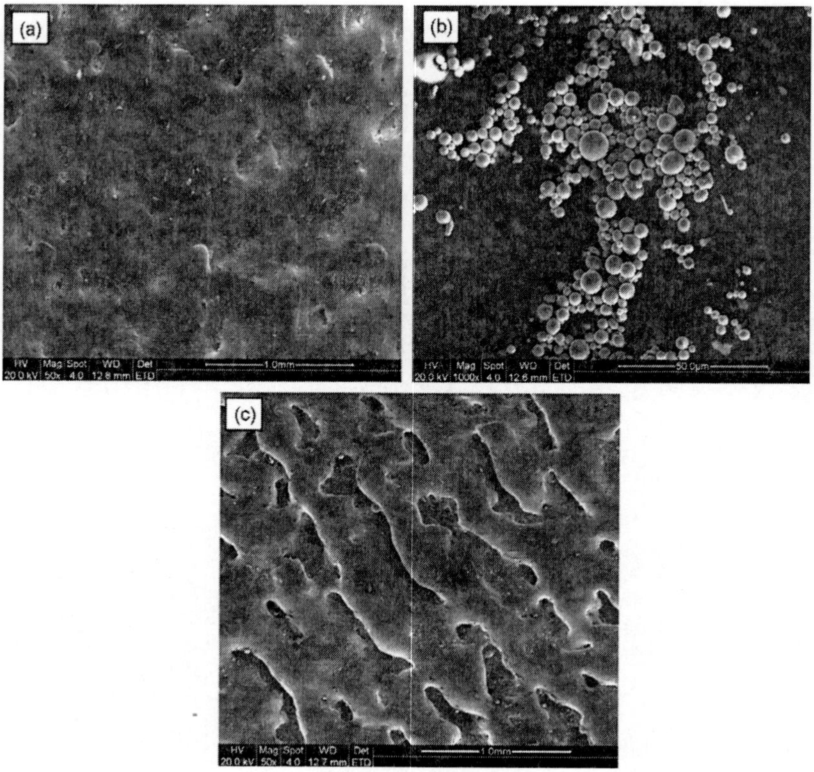

Figure 8.3: SEM Image of Specimen with different Process Parameters [3]

Above all, under a higher laser energy input, a higher temperature and attendant lower viscosity would be generated, thus, yielding a much smoother surface and denser structure. However, at a relative lower laser power of 80 W, many agglomerates were formed, between which pore channels could be observed, indicating a worse wetting ability and porous surface clearly shown in figure 8.3(c). The unfavourable flow ability under a lower laser power cannot easily urge the liquid to fill pores, incurring the occurrence of many metallic agglomerates so as to reduce the surface energy [4, 5].

Figure 4 shows the characteristic microstructures on polished cross sections of samples obtained from gas atomized powder via different scan speeds. At a relative lower scan speed of 100 mm/s, a denser SLM-produced microstructure could be obtained, as shown in figure 8.4(a). There were no apparent big sized pores in this structure, although some micro pores could be detected. As the scan speed increased to 200 mm/s, the pore quantity was enhanced and the pore size was also enlarged figure 8.4(b). At an even higher scan speed of 300 mm/s, a weaker powder bonding condition was visible, with a large amount pores formation in the structure.

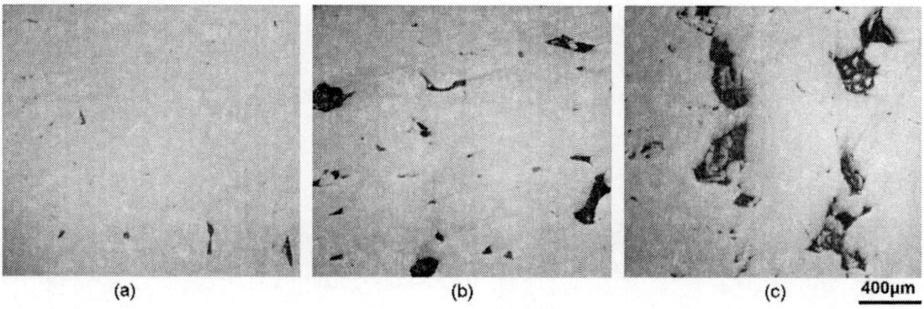

Figure 8.4: SLM-produced Samples at (a) v = 100 mm/s; (b) v = 200 mm/s; (c) v = 300 mm/s; h = 0.15mm, d = 0.05mm and P = 190W are Constants [3]

Figure 8.5 describes the characteristic microstructures on polished cross sections of samples obtained from gas atomized powder via different hatch spacing. It can be found that the hatch spacing plays a crucial role in determining the porosity, pore size and pore shape, according to different overlapping conditions of laser single scanning tracks. At a lower hatch spacing of 0.15mm, the laser scanning tracks could be overlapped tightly, thus a dense microstructure with very little micro pores was obtained is shown in figure 8.5(a). When the hatch spacing increased to 0.3mm, the pore fraction and size were enhanced and the pores exhibited an irregular shape in figure 8.5(b), due to the week overlapping ability at wider hatch spacing. With further increasing the hatch spacing to 0.4mm, neatly aligned pores in big size and regular shape came into being displayed in figure 8.5(c), since the single scanning tracks could not be overlapped at the wider hatch spacing. As the hatch spacing increased to 0.8mm, the single scanning tracks were separated in long distances, thus the extremely big pore in regular shape tended to be formed in shown in figure 8.5(d).

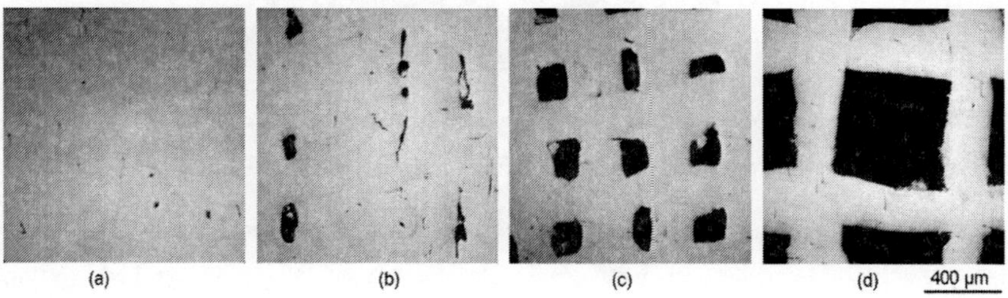

Figure 8.5: SLM-produced Samples at different Hatch Spacing; d = 0.05mm. P = 190W and v = 100mm/s are Constant [3]

Figure 8.6 shows the pore structure in polished section at different layer thicknesses and powder kinds. Figure 8.6(a) showing extremely little distributed pores for SLM of gas atomized powder at layer thickness of 0.05mm that could yield a dense structure. However, SLM at a thicker layer thickness with the same powder tended to induce many uniformly distributed pores and worse layer bonding condition as shown in figure 8.6(b). SLM of water atomized powder at layer thickness of 0.1mm was apt to induce more pores than that obtained from gas atomized powder as shown in figure 8.6(c).

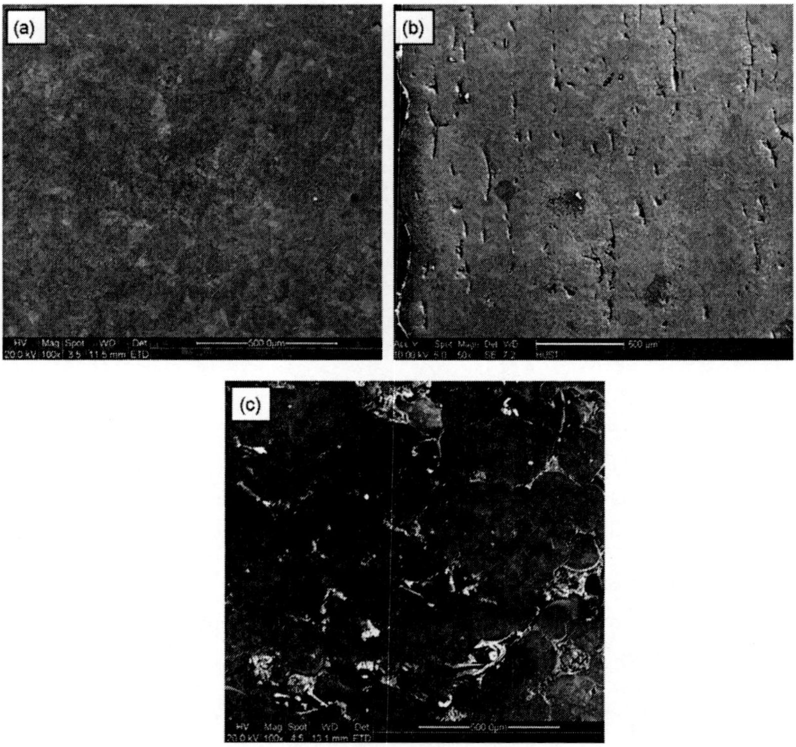

Figure 8.6: Samples at different Layer Thicknesses and Powder Kinds: (a) d = 0.05mm, gas Atomization; (b) d = 0.1mm, Gas Atomization; (c) d = 0.1mm Water Atomization [3]

The EDX analysis revealed that the water atomized powder contained more oxygen content than gas atomized powder. Oxidation could also be a vital factor for the bad wetting characteristics. The molten pool does not easily wet the surface coated with oxidation, thus, a porous melted layer would be formed. Moreover, the gas atomized powder had a spherical shape and a smaller mean particle size than water atomized powder, and thereby possessed a higher the packing density, which was in favour of SLM densification.

Laohaprapanon, et al [6] A conducted an experiment on the optimal Scanning Condition of Selective Laser Melting Processing with Stainless Steel 316L Powder gas atomized powder particles with the metallic power was filled into the substrate with 100 μm thickness. The line contour was controlled the direction using AutoFab. A 15mm single line scanning track was performed with a laser power of range 25-225 W and 50-320 mm/s scanning speed and the scan tracks are shown in figure 8.7.

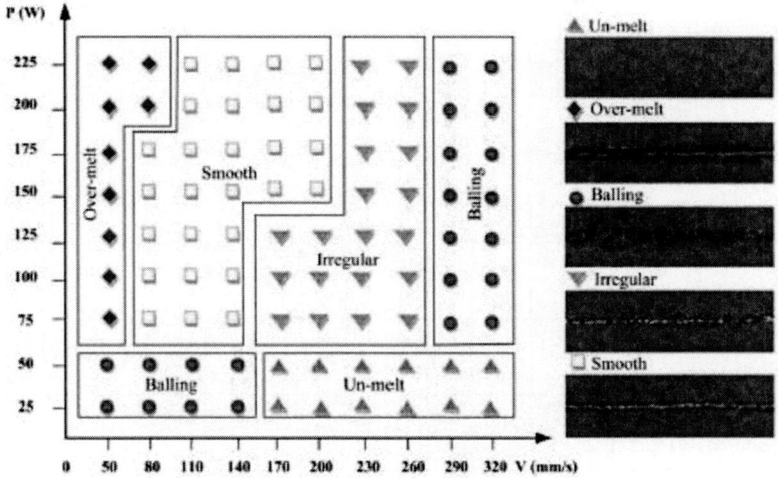

Figure 8.7: Effect of Laser Power and Scan Speeds on Substrate Appearance of Line Spacing [6]

In order to analyze the quality of each condition of single line scanning, the scan track was observed visually. The appearance of scan tracks was categorized into 5 models which were un-melted, balling, smooth, irregular, and over-melt. For multiple layers scanning, the work pieces were used to carry out the density based on the principle of Archimedes' law (ISO 3369). All work pieces were also examined the hardness using Vickers micro hardness tester. The hardness test protocol was to apply a 400 g load to the surface of work piece for 15 seconds and cross-section of parts was examined by characteristic micrograph.

The optimal condition from the single line scanning was used to build the cube for multiple layers scanning, condition that caused some of "smooth tracks" and some of "irregular tracks". Based on 2^k factorial analysis, the selected laser powers used were 75 W and 175 W and the selected scanning speeds were 80 mm/s and 200 mm/s. Therefore, eight experiments were conducted and each of experiments built two cubes to confirm repeatability. All cubes were examined density, hardness and build rate.

For multiple layers examination, the effect of scan spacing and scanning strategy were focused. Two scan spacing, 0.04 mm and 0.06 mm were under consideration. The investigation design was to construct the 10 mm cube work pieces based on cross-hatched scanning strategy [7].

Table 8.2: Characteristics of Work Piece after Processing with some of Smooth Tracks and Irregular Tracks Conditions [6]

Laser Power(W)	75W				175W			
Scan speed(mm/s)	200	200	80	80	200	200	280	80
Scan spacing(mm)	0.06	0.04	0.06	0.04	0.06	0.04	0.06	0.04
Layer thickness(mm)	0.1	0.1	0.1	0.1	0.1	0.1	0.1	0.1
Build rate(cm^3/hr)	4.32	2.88	1.73	1.15	4.32	2.88	1.73	1.15
Energy density(J/mm^3)[8]	63	94	156	234	146	219	365	547
Relative part density (%)	81.65	80.40	88.47	89.29	88.7	97.57	97.81	88.51
Micro Hardness(HV)	203±6	208±6	215±6	208±5	216±6	220±6	227±5	228±4

The processing condition was the optimal one obtained previously the line scanning experiments. The physical density analysis was based on Archimedes' law. The hardness of work pieces was examined by Micro-hardness (Vickers) and build rates were calculated as in equation 8.1.

$$B_r = L_t \cdot S_s \cdot V \qquad (8.1)$$

Where: B_r is Build rate (cm^3/h), L_t is Layer thickness (μm), S_s is Scan spacing (mm) and V is scanning speed (mm/s) [6] and the obtained results are shown in table 8.1.

From the single line scanning test, it can be noticed that the "un-melt tracks" causes where the laser power is lower than 50 W. At this stage, the energy generated from the laser is not sufficient to melt the metallic particles. Consequently, the particle individually remains without melting. The "balling tracks" occurs by two scanning conditions: Condition (A)-relatively low laser power (25- 50 W) and scanning speed ranging from (50-140 mm/s) and Condition (B)-high laser power (75- 225 W) and scanning speed above 290 mm/s. It can be obviously seen that liquid-solid mixture (melt pool) having high viscosity are consequence from Condition (A). In contrary, Condition (B) occurred at scanning speed above 290 mm/s with enough laser power deposition, the molten track is formed. During the formation, the molten track tries to reduce surface energy becoming small balling [9].

The smooth tracks can be formed by laser power between 75-225 W and scanning speed at 90-200 mm/s. The phenomena can be described as sufficient liquid stability, continuous-and smooth melted tracks without balling formation. The "irregular tracks" is caused by high scanning speed (170-260 mm/s) and high laser power (75-150 W). The appearance of track is generally distorted and un-continued. This is subsequent from reduction of surface energy and instability of liquid. The "over-melt tracks" is normally overheating, oxidation and tracks widening which subsequently presented as large melt pool. This generally occurs at low scan speed and high laser power. From the results of the multiple layers scanning as table 8.2, it can be seen that the density of the work pieces were influenced by the input energy density. Higher energy density leads the higher part density as well as increasing the hardness.

Figure 8.8(a-c) demonstrates a surface cross-section of the parts produced from 3 different energy densities. Under the different treatment conditions, the values of relative density and hardness of the parts were similar and the maximum relative density and hardness were 98.51% and 228±4 HV at the highest energy input 547 J/cm^3. However, considering the build rate, figure 8.8(a) condition has the highest value 2.88 cm^3/compared to condition b and c. Based on the results of relative density, hardness and build rate, treatment condition (a) was used as the optimum condition in the present study and concluded that the optimal conditions for processing stainless steel 316L using SLM technique should be the conditions within the smooth tracks region. The processing parameters including laser power, scanning speed and scan spacing have directly influence with the part density and hardness. The optimal condition of multilayer part is P 175 W, V 200 mm/s and Ss 0.04 mm.

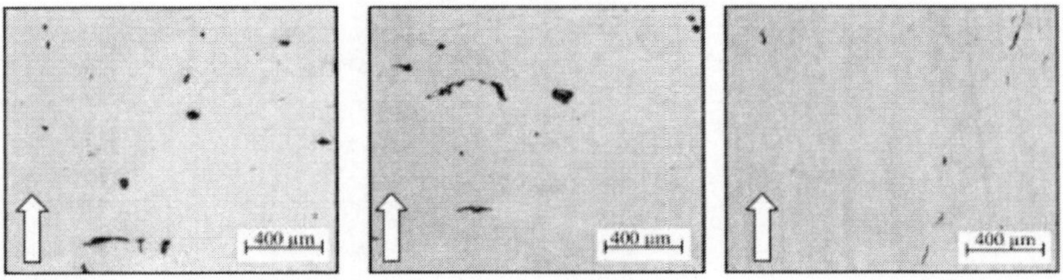

(a) D = 97.57% (b) D = 97.81% (c) D = 98.51%

Figure 8.8: Cross Sections of SS 316L Parts with different Condition [6]

8.3. Biocompatibility of Stainless Steel with Respect to Scanning Strategy

Shang, Y, et al [10] and his team deliberated the effects of scanning speed on in vitro biocompatibility of 316L stainless steel with 316L SS powder size range from 24 to 59 μm, the mean particle size 38 μm, and manufactured via gas atomization and SEM image shows characteristic morphologies of 316L SS as shown in figure 8.9. The chemical composition of material is given in table 8.3.

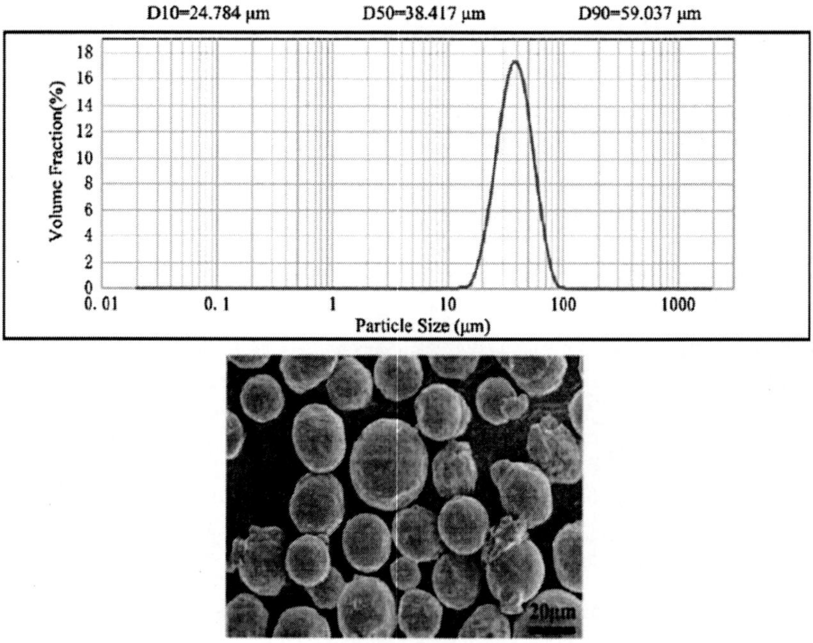

Figure 8.9: Size Distribution and SEM Image of 316L SS Powder Particles [10]

Table 8.3: Composition of SS316L (Wt%) [10]

	C	Cr	Ni	Mo	Si	Mn	P	S	Fe
Grade SS316L	0.030	16.97	13.11	2.35	0.75	2.00	0.025	0.010	Bal
ASTM F55	0.030	16-18	10-14	2-3	0.75	≤ 1.06	-	-	Bal

During SLM processing, a range of process parameters, such as scanning speed, laser power, hatch spacing and environmental parameters, can be changed to achieve the optimum material properties. In our experiments, the processing parameters are as follows: scanning speed (v) of

800, 900, and 1083 mm s−1, laser power (P) of 195 W, hatch spacing (h) of 0.09 mm and layer thickness (d) of 0.02 mm. In order to reduce the errors caused by the experimental operation and data, three repeats of the samples are fabricated for each setting, namely 9 cylindrically shaped samples 10 mm diameter and 10 mm thickness as shown in figure 8.10 are manufactured by EOSM280.

Figure 8.10: Cylindrical Shaped Samples [10]

For Cytotoxicity test L929 fibroblast cell line is adopted to evaluate the cytotoxicity of experimental materials, which has been demonstrated sensitivity to extractable cytotoxic articles and obtained from ATCC (American Type Culture Collection). The cytotoxicity tests are performed by indirect contact and the 9 cylindrically shaped samples are sterilized by irradiation before experiment. Extracts are prepared using minimum essential medium (MEM) with addition 10% fetal bovine serum (FBS) as the extraction medium in the ratio of 0.2 g: 1 ml in a humidified atmosphere(5% CO_2, 37 °C) for 24 h. In addition, high density polyethylene as negative control, zinc diethyldithiocarbamate as positive control and MEM medium with addition 10% FBS as blank control. L929 cells are cultured in MEM medium at 37 °C in a humidified atmosphere of 5% CO_2, and then digested by 0.25% trypsin containing EDTA to get single cell suspension. 1×10^5 cells ml^{-1} suspension is obtained by centrifuging and re-dispersing. The suspended cells are dispensed at 100 μl per well in 96-well plate, and then cultured in cell incubator (5% CO_2, 37 °C, >90% humidity). After cells growing and forming a monolayer, original culture medium can be discarded. Then 96-well plates are treated with 100 μl extracts of test articles, control article, negative article and positive article, respectively. After cells are incubated for 24 h, the cell morphology is observed under an optical microscope. Cell proliferation is examined by the MTT assay.

L929 fibroblast cells are cultured in the extract with 50 μl aliquot of MTT (1 mg ml−1) for 2 h at 37 °C in a humidified atmosphere of 5% CO_2. Then, the medium in each well can be discarded and 100 μl 99.9% isopropanol is added to each well to make the cell layer suspend. The spectrophotometric absorbance of samples is measured by Power Wave XS Micro plate Reader (ESZ4007) at 570 nm with a reference wavelength of 650 nm. All cell experimental investigations are carried out in quintuplicate.

Hemolytic test is carried out according to ISO 10993.4. Diluted rabbit blood is prepared with 0.9% sodium chloride water (ratio of 4:5 by volume).The 9 cylindrically shaped samples are washed three times in sterile ultra-pure water. Experimental samples are immersed in 0.9% sodium chloride water at a ratio of 0.2 g:1 ml. Mixture of diluted blood and the extracts of experimental samples (0.2 ml diluted rabbit blood: 10 ml extract) is incubated for 60 min. Similarly, negative control in which 10 ml of 0.9% sodium chloride water is added to 0.2 ml of diluted rabbit blood and positive control in which 10 ml of distilled water is added to 0.2 ml of diluted rabbit blood are performed in this test. All experiments are carried out in quintuplicate to reduce accidental error. The mixed solution in each tube is centrifuged at 2500 rpm for 5 min. The supernatant is examined at a wavelength of 545 nm and recorded by a spectrophotometer (7230G, Shanghai Qili). The hemolytic ratio is expressed as a percentage and is calculated according to the following formula:

$$\text{Hemolytic rate} = \frac{A-C}{C-B} \times 100\ \% \tag{8.2}$$

In vitro cytotoxicity evaluation Figure 8.11 depicts the morphology of L929 cells after culture for 24 h in different extracts.

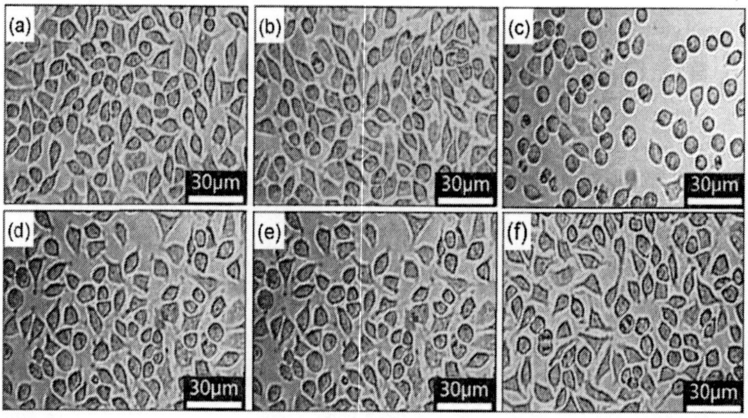

Figure 8.11: Morphology of L929 Cells after Culture in different Extracts [10]

The morphology of normal and healthy cells is as shown in Fig. 8.11(b), which is similar to that of the blank control group in Fig. 4a, whereas abnormal morphology (round cells) is observed in the positive control group, as shown in Fig. 8.11c. Figure 8.11(a–c) indicates the experiments of in vitro cytotoxicity are valid. Figure 8.11(d–f) shows the discrete intracytoplasmatic granules observed after 24 h of cells incubation in extracts and the difference of cell morphology at different scanning speed. At the scanning speed of 800 mm s^{-1}, the ratio of round cells, about 25%.

For the case of 900 mm s^{-1} scanning speed, the ratio of round cells, about 10%, and the ratio is about 16% at 1083mm s^{-1}. The experimental results demonstrate that (1) the scanning speed strongly affects the L929 cell morphology and (2) the maximum ratio of normal and healthy cells is obtained at 900 mm s^{-1} scanning speed as shown in figure 8.12.

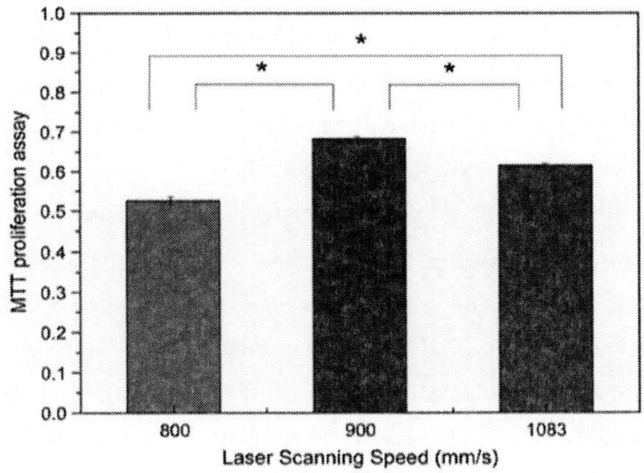

Figure 8.12: MTT Proliferation Assay of Cells Grown for 24 h in the Test Article Extracts at Three different Scanning Speeds [10]

Figure 8.12 illustrates the MTT proliferation assay of cells grown in the test article extracts at three different scanning speeds. For the 900 mm s^{-1} scanning speed case, the MTT proliferation (0.6826 ± 0.000559) is significantly higher than those of the other groups ($P < 0.05$), while the MTT proliferation (0.5256 ± 0.000622) decreases dramatically at the scanning speed of 800mm s^{-1}. The scanning speed of 1083 mm s^{-1} part shows a median MTT proliferation (0.6150 ± 0.000450) among the three groups. The cytotoxicity test data shows that the scanning speed has strong effect on the in vitro biocompatibility. Clinical experience has revealed that metal toxic ions (Ni^{2+}、 Cr^{3+}、 Mo^{6+}) released from 316L SS implants, lead to in vitro cell death and loss of the cells integrity which will harm cellular metabolism [11].

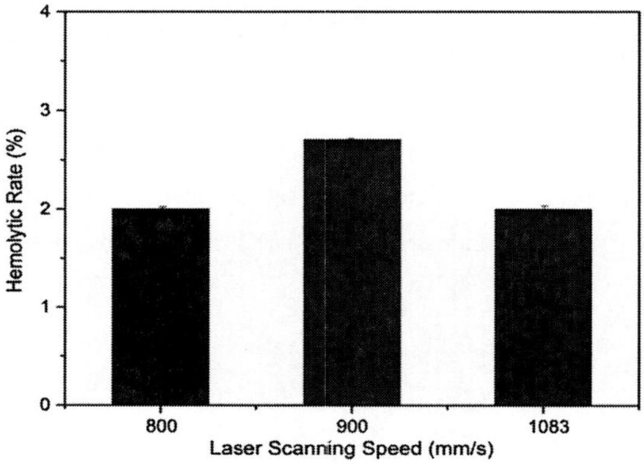

Figure 8.13: Hemolytic Rate of SLM 316L SS Parts at Three different Scanning Speeds [10]

In vitro hemolytic evaluation in figure 8.13 describes the hemolytic rate of SLM 316L SS parts at three different scanning speeds. No statistical difference is observed among the three groups. Three groups have a similar and extremely low ratio of hemolysis, which is no more than 5%. The hemolytic data indicates that scanning speed has no significant effect on hemolytic properties. Metal toxic ions (Ni^{2+}、Cr^{3+}、Mo^{6+}) released from 316L SS implants have no significant effect on hemolysis in this study. SLM is a promising way for SS316L implants manufacture and concluded that the scanning speed has strong effects on in vitro cytotoxicity of SLM316L SS parts and no significant effects on the hemolysis. For the 900 mm/s scanning speed case, the MTT cells proliferation is significantly higher than those of the other groups ($P < 0.05$), and the maximum ratio of normal and healthy cells is obtained (the ratio of round cells is around 10%).

8.4. Sustainability of Steel Powders

Geenen, K., et al [12] investigated the microstructure, mechanical, and tribological properties of M3:2 high-speed steel processed by selective laser melting, hot-isostatic pressing, and casting and usage of powders is investigated in terms of sustainable manufacturing. During SLM processing, oxygen can be absorbed by the heated powder. If this partially oxidized powder is used again for SLM densification of further components, the increased oxygen content can adversely affect the remelting behaviour, the resulting microstructure, and ultimately the material properties. For this reason, the reuse of steel powder after several SLM cycles has been studied. The reuse of powder that was previously processed by selective laser melting is discussed controversially in the respective specialist literature [13, 14]. Reusability of the powder is generally not excluded in principle. In this work, the M3:2 high-speed steel powders were closely investigated with respect to powder reusability both before and after several processing cycles. It should be noted that the used powder is mechanically sieved after each SLM cycle and thus divided into fractions of<63 µm (reused powder) and>63 µm (sorted residues). The particle morphology, particle size distribution, flow behaviour, apparent or packing density, and oxygen content were each investigated with the sieved powder fraction of< 63 µm. Micrographs of the sieved M3:2 high-speed steel powder particles are shown in figure 8.14.

Figure 8.14: Micrographs of M3:2 HSS Powder in the Initial State [12]

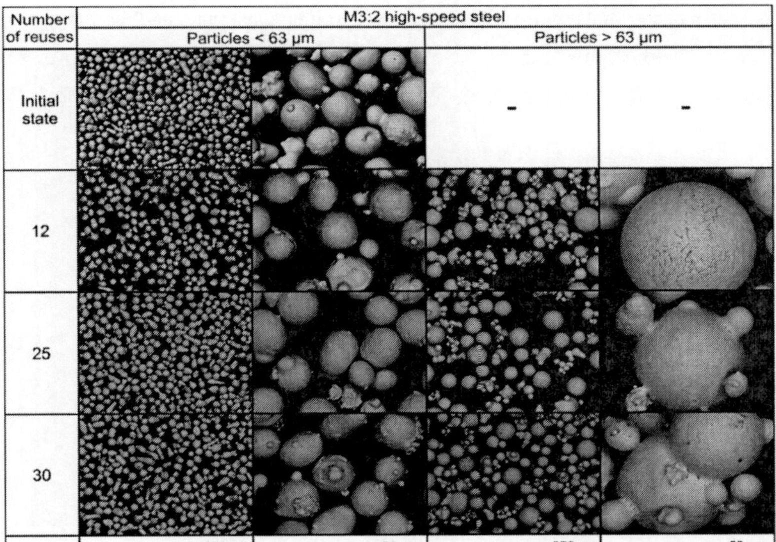

Figure 8.15: Micrograph of Sieved M3:2 High-speed Steel Powder Particles [12]

The morphology as shown in table 8.4 indicates the powder particles with a size less than 63 µm corresponds to the morphology of the steel powder in its initial state, as presented in figure

8.14(a). However, the micrographs in figure 8.15 indicate splashes and the formation of bigger agglomerates of powder particles larger than 63 µm.

Table 8.4: Particle Analysis of the M3:2 Steel Powder in the Initial and in the Reused States (After 30 Cycles) [12]

Sample	**Flow rate (s/ 50g)**	**Apparent density (g/mm)**	**Packing density**	**Particle diameter (µm)**		
				d_{10}	d_{50}	d_{90}
M3:2 powder in initial state	13.46 ± 0.04	4.43	0.55	31	47	68
M3:2 powder after 30 cycles	14.26 ± 0.04	4.52	0.56	38	51	68

In the latter case, the powder is characterized by the formation of distinct sinter bridges caused by partial interaction between the laser beam and the powder particles during SLM process. The findings underline the necessity for (mechanical) sieving of the powder because the splashes or agglomerated particles could, for example, hamper the flow behaviour. The particle size distribution was measured after 30 reuses of the high-speed steel powder in the SLM process and shows good agreement with the initial state, as can be seen in figure 8.16.

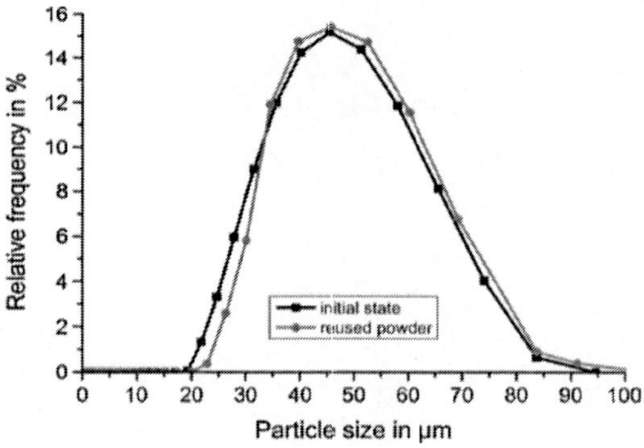

Figure 8.16: Particle Size Distribution of the M3:2 High-speed Steel in the Initial and Reused States (After 30 Cycles) [12]

In addition, the multiply reused steel powder exhibits nearly the same particle diameters, flow rate, and apparent or packing density with respect to the steel powder in the initial state as shown in table 8.4. Hence, it can be concluded that multiple reuse of the high-speed steel powder for SLM processing is possible without any significant alteration of the particle properties.

However, the powders must be classified after each SLM cycle. Furthermore, the oxygen content of the high-speed steel powder in its initial state as well as after different numbers of processing cycles was measured and compared to those of cast and hot-formed as well as HIP-densified samples which is indicated in table 8.5.

Table 8.5: Comparison of M3: 2 High Speed Steel Powders [12]

Designation/Number of cycles or reuses	**O content of particle size < 63 µm [ppm]**	**O content of SLM densified samples [ppm]**	**O content of reference sample [ppm]**
Cast	-	-	52 ± 3
HIP-densified	-	-	207 ± 13
Initial state	187 ± 7	205 ± 7	-
After 5 cycles	225 ± 2	224 ± 1	-
After 12 cycles	239 ± 4	246 ±0	-
After 25 cycles	239 ± 2	268 ± 1	-
After 30 cycles	243 ± 2	277 v 4	-

The steel powder in the initial state has an average oxygen content of about 190 ppm. This is higher than the measured oxygen content of specimens in the cast and hot-formed condition (≈52 ppm), but lower than the oxygen content of the HIP densified specimens (≈210 ppm). However, the higher oxygen content in the HIP specimens can be attributed to the higher oxygen content of the powder batch densified by HIP. After each reuse of the sieved powder (smaller than 63 µm) for further processing by SLM, the oxygen content of the steel powder continuously rises up to 225 ppm after 5 SLM cycles and 243 ppm after 30 cycles. The SLM-densified specimens show the same behaviour for the increase in oxygen as shown in table 8.5, but the oxygen content measured in the SLM specimens is up to 35 ppm higher than in the steel powder used for the SLM-densification cycle. In contrast to the oxygen content, the concentration of all other elements in the residual high-speed steel powder and SLM specimens, such as carbon, remained almost unchanged. In previous studies, the oxygen uptake by the powder and the SLM-densified specimen of 316 L austenitic steel and X40CrMoV5-1 hot-work tool steel was discussed in a different manner [15, 16].

On the one hand, it was described that the oxygen is absorbed by the steel powder surfaces during the decanting and mechanical sieving process, which was performed under a normal atmosphere and not using a protective gas. In this regard, the oxygen absorption can be described as a regressive trend owing to saturated uptake of oxygen by the steel powder surface, meaning that the upper limit of the oxygen content is a function of the specific surface area of the respective powder. On the other hand, the interaction between the powder particles and the laser beam leads to a high heat input into the particles that promotes oxygen uptake. Thus, the presented results indicate that the processing atmosphere of the SLM building chamber (residual oxygen content: ≈0.3 vol%) can be viewed as the primary source of the higher oxygen content because most of the oxygen is absorbed by the laser-heated material [15, 16]. In short, it can be concluded that the M3:2 high-speed steel powders can be reused, as in practice, for economic and environmental benefits. However, it should be considered that the powder and the SLM-densified specimens can lose quality as a result of the oxygen uptake after each SLM cycle and this might negatively affect the various material properties due to the formation of oxide inclusions in the microstructure.

8.5. Mechanical Properties of SLMed Parts

8.5.1. Hardness

Alsalla, H. et al [17] deliberated the effect of build orientation on the mechanical properties along with surface quality, microstructure of SS316L powders with different orientation as shown in figure 8.17 were produced with a laser power of 180 W, a scan speed of 1600 mm/s, a layer thickness of 30 mm, and a scanning laser spot with a 75 mm diameter.

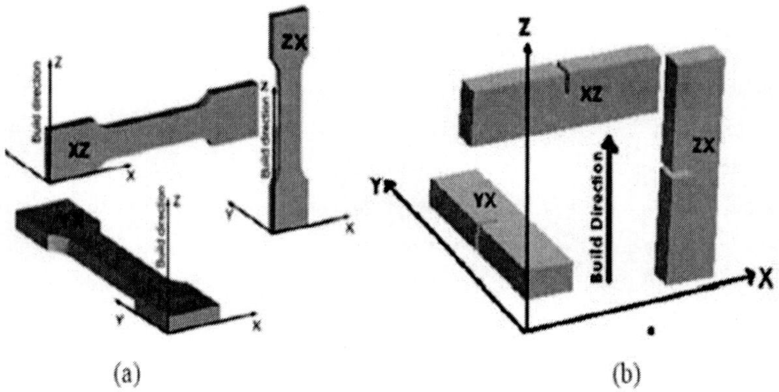

Figure 8.17: Samples build in different Directions [17]

The samples were analysed for the actual surface at different building strategies by surface profilometer (Talyscan-150, Taylor Hobson Ltd), the data were collected through two measurements, measured on the parallel and perpendicular to the building direction. The average values of Ra were obtained from both parallel and perpendicular measurements related to fracture toughness.

Fracture toughness test according to ASTM E399-9 results (three point test through compression to failure), presented in table 8.6, were calculated according to the maximum flexural load by concentrating the stress in the area of minimum cross section of as built samples made in different build directions. The fracture toughness varied between the two values 145.5 MPa $m^{(1/2)}$ in the zx direction (standard deviation ±1.1) and 176 MPa $m^{(1/2)}$, in the xz direction (standard deviation ±0.9) and a situated low value when compared to the conventional method of austenitic stainless steel grade 316L, from 112 up to 278 MPa $m^{(1/2)}$, and 316L annealed stainless steel at 210 MPa $m^{(1/2)}$. These low values should be caused by the cracks, pores and voids in the parts produced, as well as the residual stress that remains in the 316L parts after the SLM process, are other reasons that affect the mechanical properties of materials, especially fracture toughness.

Table 8.6: Fracture Toughness Value for different Build Directions [17]

As built sample	Fracture toughness (MPa $m^{(1/2)}$)
YX	152.6±1.16
XZ	176±0.9
ZX	145.5±1.1

The differences in the fracture toughness values obtained in this study with different build directions had been anticipated. The tensile test result also showed that the samples built in the xz direction recorded the highest values of UTS and elongation, meaning that there is a larger area under the stress strain curve, i.e., the material absorbs more energy before failure. These combined values enhance the fracture toughness of parts built in the xz direction because of the fast cooling rate and the resulting dendritic grain structure. On the other hand, parts built in the yx and zx directions presented slightly lower values of fracture toughness than those built in the xz direction because of the more prevalent and larger pores and defects, especially in the samples built in the zx direction, which were subjected to loading perpendicular to the build direction. The results in table 8.6 reveal that the build orientation has a strong effect upon properties. For example, fracture toughness was smallest in the samples built in the z direction because of the pores, voids and cracks that are present at the edge of the parts.

The Vickers hardness test for as build samples built in different build orientations was found to demonstrate no significant difference between 226 HV in the xy samples and 221HV in the xz and zx samples as shown in figure 8.18, with a standard deviation of 2 HV. The xz build records the lowest value of hardness, which means that the samples that were built in the xz direction have more ductility than those built in other directions that can be noticed, while the zx direction build also has a low hardness due to the porosity that is revealed in this build direction, see table 8.7. The result presented in this study is similar to those found in conventional products that are made of this alloy, and in a previous study of SLM 316L.

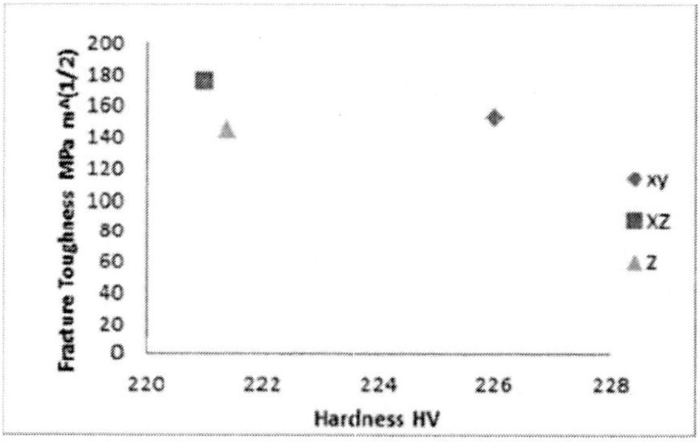

Figure 8.18: Fracture Toughness of 316L Stainless Steel Plotted against the Hardness HV [10]

Table 8.7: Vickers Hardness Results of SS316L [17]

Build direction	Vickers hardness ±HV	± Stnd Dev HV
YX	226	2
XZ	221	2
ZX	221	2

Hardness is the term used to describe the amount of energy required to permanently deform (stretch, bend, compress, etc.) a material. Toughness describes the energy required for fracture. It is often the case that hard materials are not tough, and vice versa (Osakada & Shiomi, 2006, Leskovšek, Ule, & Liščić, 2002). In this study, the xz samples showed the least hardness, as shown in figure 8.15, but they also demonstrated the highest fracture toughness, as well as the highest strength and ductility values at 695 MPa and 41 %, respectively.

8.5.2. Tensile Strength

Alsalla, H. [17] deliberated the effect of build orientation on the mechanical properties such as tensile strength along with surface quality, microstructure of SS316L powders with different orientation as discussed in section 8.5.1.

Specimens as shown in figure 8.16 were used and pulled until they fractured for the tensile testing according to ASTM E8. Figure 8.16 reveals the tensile stress-strain curves that were calculated for the yx, xz and zx build direction. The elastic slope in this figure is usually named the Apparent Young's Modulus. The total elongation to rupture, the ultimate and yield strength were determined after examination of the plastic and elastic parts of the curves; see table 8.8 and figure 8.19.

Table 8.8: Tensile Test Results for different Build Directions [12]

As built sample	UTS(MPa)	Yield strength(MPa)	Elongation (%)
YX	668±3	397±3	37±1
XZ	695±3	423±3	41±1.9
ZX	564±3	387±3	35±0.6
Cast material	515	205	60

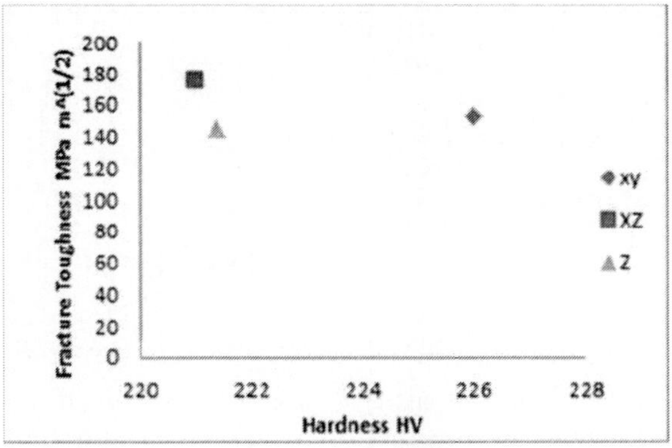

Figure 8.19: Stress-strain Curves for the SLM of SS 316L Built in different Directions [17]

The total elongations were found to be between 35% and 41% and these results corroborate with the results presented in previous studies that were carried out by John and Yang et al and the ultimate tensile strength (UTS) recorded a large difference from 564 MPa for the zx samples, to 695 MPa for the xz samples (standard deviation 3 MPa). The lowest value of UTS was recorded in the z samples, because of the number of irregular pores and the presence of a weak region between the scanning tracks and the melting layers where all the fractures had occurred.

On the other hand, the yield strength varied from 387 in the zx samples to 423 MPa in the xz samples, and was situated above the previous research results of 261MPa. This enhancement in the tensile properties was due to the proper selection of the process parameters.

8.6. Tribological Behaviour of SLMed Parts

8.6.1. Friction and Wear

Zhu, Y., et al [18] investigated the tribology of selective laser melting processed stainless steel 316L under lubricated conditions with the samples are shown in figure 8.20. Lower samples were made of a soft material H68 brass and a hard material 38CrMoAl that has been hardened to a depth of 0.5mm using ion-nitriding technology.

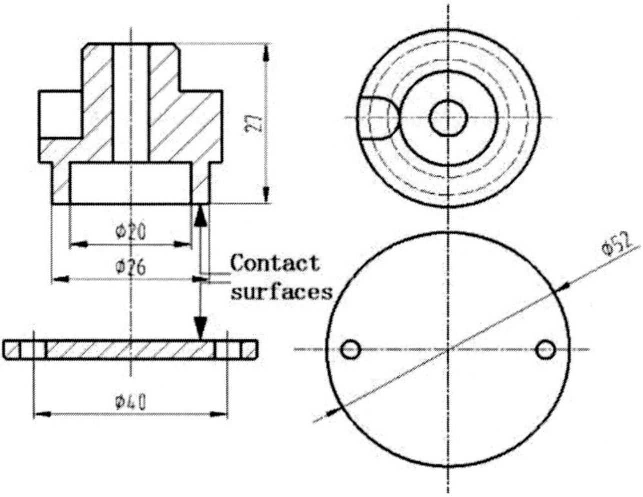

Figure 8.20: Dimensions of Upper and Lower Specimens [18]

The hardness of the brass and 38CrMoAl was approximately 110 HV and 1001 HV, respectively. Hardness tests were performed by Vickers Hardness tester with a 1kg load. Mean values were obtained through six measurements on one specimen. Lower samples were produced using the traditional process.

The upper samples were made of 316L stainless steel and produced through both SLM and the traditional process in order to have a comparison. SLM was performed using a Renishaw AM250 with process parameters, an argon gas protection system, and a process control system.

Each powder layer was scanned by the laser beam and samples were built in accordance with the parameters listed in table 8.9. 316L stainless steel powders were provided by Renishaw plc., the size of which ranged from 15 μm to 45 μm with a mean diameter of 30 μm.

Table 8.9: Process Parameter for Sample Preparation [18]

Layer thickness	Spot size	Hatch space	Laser power	Point distance	Exposure time	Scan speed
30μm	70 μm	110 μm	200 W	70 μm	100 μs	700m/s

The contacting surfaces of all TP (Traditionally Processed) samples were finished by coarse grinding and polishing. The root mean square (RMS) value Ra of samples is 0.8 and 0.02 μm respectively. SLM components are known to have a relative rough surface finish (approximately Ra = 5 μm is the best it can achieve), which is far from the requirements of real engineering application. Therefore, the SLM samples were also employed by coarse grinding and polishing to a roughness of Ra = 0.8 and 0.02 μm respectively. This made it possible to compare the way in which different manufacturing processes affect the tribological performance at two surface finishes. The surface topography of upper and lower samples was analysed.

The rotating speed of the upper specimen was 10 rpm throughout the tests. Each test lasted for 30 min, equivalent to a sliding distance of 21.7 m. The sliding distance was calculated by the rotation speed and the siding radius here the sliding radius is the mean value of the inner radius and the outer radius. The force applied was 3300N in order to obtain a contact pressure of 15MPa. The contact pressure used here is a very typical value in the port plate-cylinder barrel contact according to Wang [19]. Quite low speed was used because the efficiency of a piston pump is low at the low speed due to the failure of the oil film. The property of the oil is listed in table 8.10.

Table 8.10: Property of Hydraulic Oil Used in Tests [18]

Property	**Values**
Kinematic viscosity(mm^2 /s at 40 °C)	48.2
Viscosity index	96.3
Flash point(°C)	196
Pour point (°C)	-16

The contact conditions were set to simulate the port plate—cylinder barrel contact in an axial piston pump under low speeds, since pump efficiency at low speeds is low due to insufficient lubrication. The hydraulic oil used in the tests was L-HM 46 anti-wear hydraulic oil, which was one of the most widely used hydraulic oils in China. The flow rate of the oil is 2 ml/s in order to keep the contact lubricated and prevent the oil from overheating. Upper and lower samples with the same manufacturing process and surface finish were tested as one pair. All specimens were ultrasonically cleaned for 10 min with acetone before and after testing. Each test was repeated three times. For the contact between stainless steel and brass, the wear was measured by weighing the brass sample before and after the tests, since brass is much softer and was subjected to most of the mass loss. For contact between stainless steel and 38CrMoAl, wear was measured by weighing the 316L stainless steel sample, because 38CrMoAl is much harder.

Tests were performed on a ring-on-disc rig MMU-10 which was used for a port plate study [20]. The upper sample was mounted on the main shaft driven by a servo motor. The main shaft rotated at speeds ranging from 0-3000 rpm.

The force was applied to the lower sample by a load cell through a hydraulic cylinder which raises the normal force to contact surfaces between the port plate and the cylinder barrel. In this way, the rig creates a sliding contact which can measure the coefficient of friction and wear rate between the port plate and cylinder barrel.

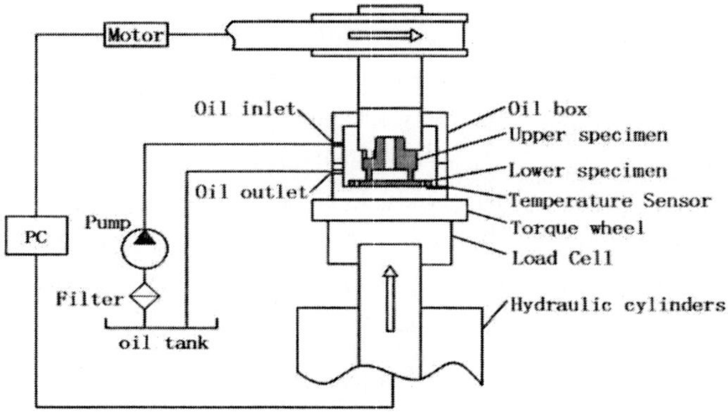

Figure 8.21: Schematic Layout of End Face Friction Tester [18]

The maximum load applied was 10 kN. The upper and the lower samples were sealed in an oil box through which oil flowed in and out. The oil box was connected to an oil tank by pipes and a filter. The flow rate of the oil from the oil tank was controlled by a pump. A schematic plot of the tribometer is shown in figure 8.21. The friction torque was measured by the torque wheel and the temperature of the lubricant was measured by a temperature sensor. The friction coefficient μ is calculated by the readings from the torque wheel T, the radius of the torque wheel r and the load cell FN, as given by equation 8.3. A computer was used to acquire data and control the load, speed and testing time. The data was recorded every second. The accuracy of the machine is ±5%.

$$\mu = T / (F_N \times r) \tag{8.3}$$

For the contact between 316L stainless steel and brass, the coefficient of friction ranged from 0.12 to 0.15 (see figure 8.22), G stands for the ground sample & p stands for polished sample which is somewhat higher than those found in a previous study in which 38CrMoAl and 42CrMo were used as the upper samples [20].

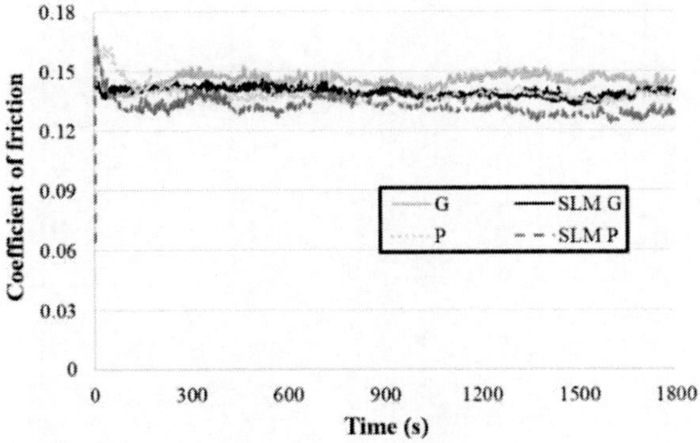

Figure 8.22: Coefficient of Friction versus Time for the Contact between SS 316L and Brass [18]

The polished surfaces gave lower coefficients of friction than ground surfaces regardless of the manufacturing process. The coefficients of friction for the SLM samples were lower than those for the TP samples. However, the differences in friction were not significant. Wear results showed the same trends as friction results did, as shown in figure 8.23. The polished SLM sample had the lowest wear rate.

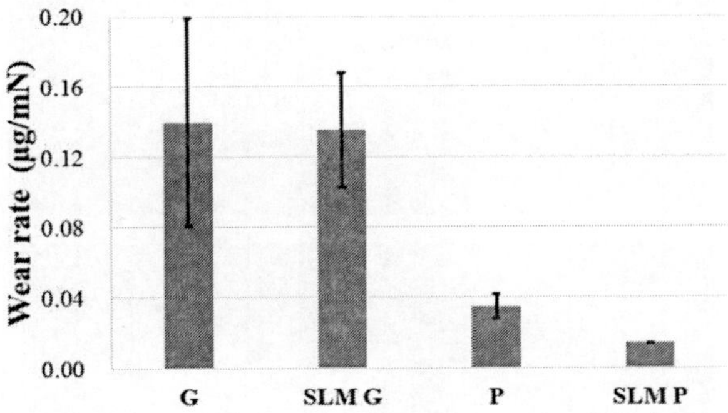

Figure 8.23: Wear Rate Measured from the Brass Samples [18]

The surface topography of samples were examined using CLSM after the tests. Brass films or particles were transferred to the 316L stainless steel samples with different morphologies, as shown in figures 8.41 and 8.25. For ground surfaces, the brass was transferred with large particles while the brass particles were much smaller on polished surfaces. The wear mechanism was previously investigated by Zhu et al. [20]. When the sliding began, the contact produced many fine particles on both the ground and polished 12 surfaces as a result of ploughing process, in which abrasive wear dominates.

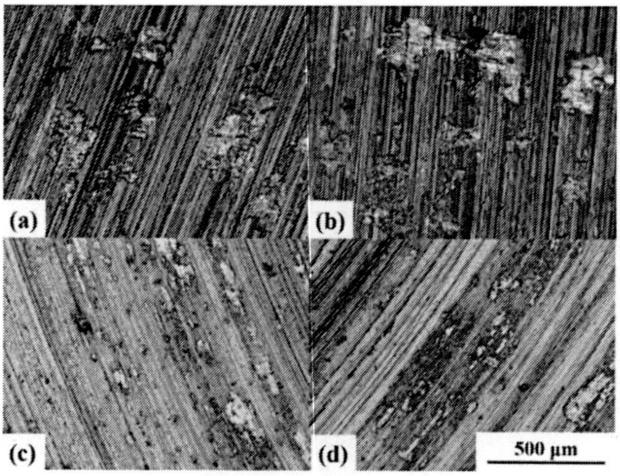

Figure 8.24: CLSM Images of Surface Topography of SS 316L Sample [18]

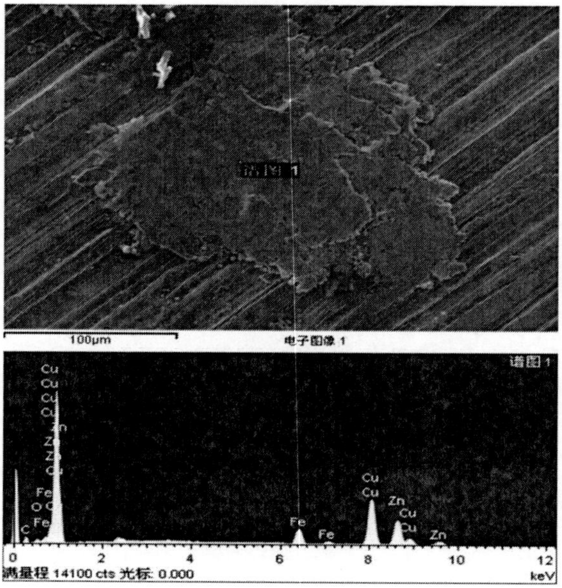

Figure 8.25: Energy Dispersive Spectrometer (EDS) Detects Brass Transfer from a SEM Image [18]

Since the difference in hardness between 316L stainless steel and brass is not as large as that between 38CrMoAl and brass [20], some fine particles were transferred to both ground and polished surfaces. The polished surfaces became smoother which prevented further abrasive wear. Simultaneously, adhesive wear appeared because part of the contact became brass-brass contact. The ground surfaces were too rough thus the ploughing continued. Small brass particles accumulated becoming large particles which increased wear on ground surfaces more than polished surfaces. The SLM processed samples showed no difference in wear mechanisms compared to the TP samples. Since 316L stainless steel came in contact with a soft material, the brass surface suffered most of the deformation. No obvious plastic deformation can be found on the subsurface as shown in figure 8.26.

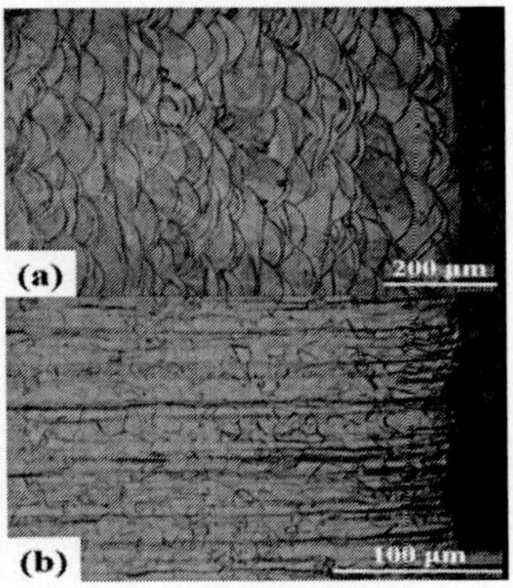

Figure 8.26: CLSM Images Showing Contacting Surface of 316L Stainless Steel: (a) SLM Sample; (b) TP Sample [18]

For the contact between 316L stainless steel and 38CrMoAl, the difference in coefficient of friction was large, as shown in figure 8.27. All TP samples and the polished SLM sample had a coefficient of friction of approximately 0.3, which is much higher than the stainless steel—brass contact had. However, the coefficient of friction for the ground SLM sample decreased as time increased until it stabilized at approximately 0.13.

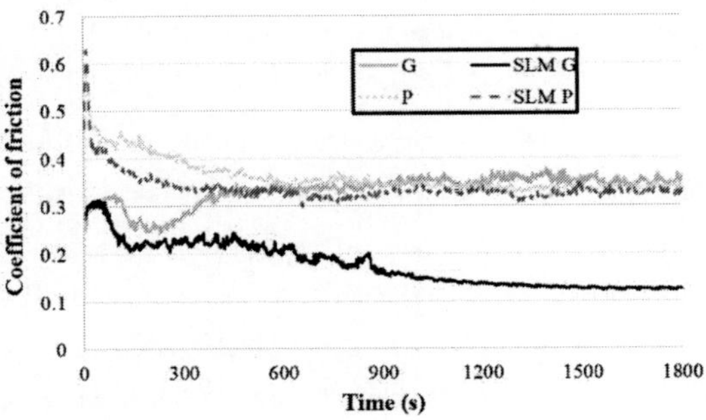

Figure 8.27: Coefficient of Friction versus Time for the Contact between SS 316L and 38CrMoAl [18]

It was noted that during the first 900 s, the variation in friction was significant. Wear results shown in figure 8.28 do not follow this friction trend. The wear rate of the ground SLM sample was half that of the ground TP sample, while was close to the wear rate of the polished TP sample. Moreover, the wear rate of the polished SLM sample was the highest. The surface topographies of the polished samples and the ground TP sample were similar. Both ploughing and adhesive wear marks can be found on the surface. In addition, both polished surfaces showed deep craters or scratches as shown in figure 8.29.

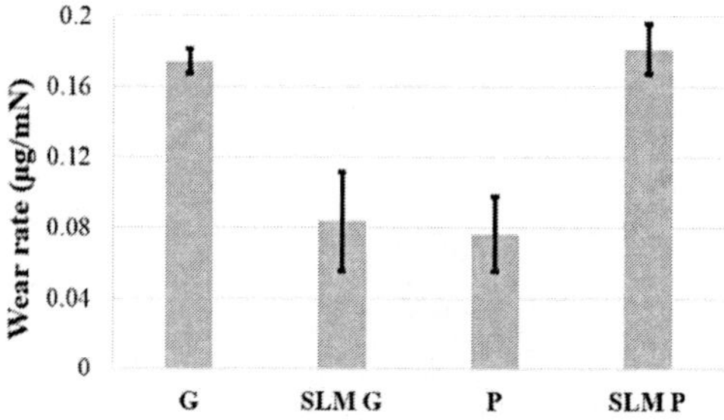

Figure 8.28: Wear Rate Measured from SS316L [18]

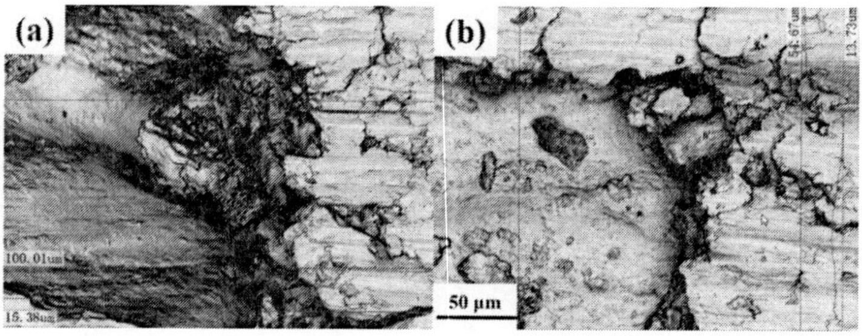

Figure 8.29: CLSM Images of Surface Topography of (a) Polished SLM Sample; (b) Polished TP Sample [18]

However, the craters had different morphologies on their surfaces. The crater on the SLM sample is approximately 85 μm, which is much deeper than that on the TP sample (approximately 40 μm). Moreover, the crater on the SLM sample was very smooth without any wear marks, which seems to be caused by shell off. The crater on the TP sample was caused by hard particles or peaks from the 38CrMoAl sample which was significantly harder than 316L stainless steel. The cross section was also observed by a CLSM shown in figure 8.30. Since the 316L stainless steel came in contact with a much harder material, most of the deformation was subjected to 15 the surface of the upper sample. The plastic deformation layer was uniformly distributed on the TP sample and the thickness was approximately 40 μm. However, the plastic deformation layer on the SLM sample ranged from approximately 20 μm to more than 60 μm. Moreover, some cracks were also observed below the surface. These cracks may result in material shell off.

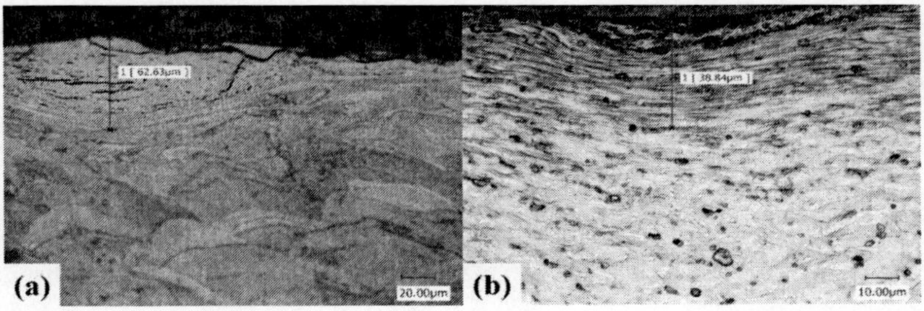

Figure 8.30: CLSM Images of Cross Section of Polished Surface of (a) SLM Sample and (b) TP Sample [18]

The ground SLM sample shows low friction and wear. The surface topography after the tests was much smoother than ground samples, as shown in figure 8.31. No deep craters were found. The EDS results in figure 8.29 show the presence of silicon, indicating material was transferred from the 38CrMoAl disc to the 316L stainless steel sample.

Moreover, the oxygen was detected in different spots (figures 8.32 a & b) on the ground SLM sample but was not found on the polished SLM sample (figure 8.32 c). It is widely accepted that an oxide layer that formed on the surface could avoid direct metal-metal contact, thus reducing friction and wear [21, 22].

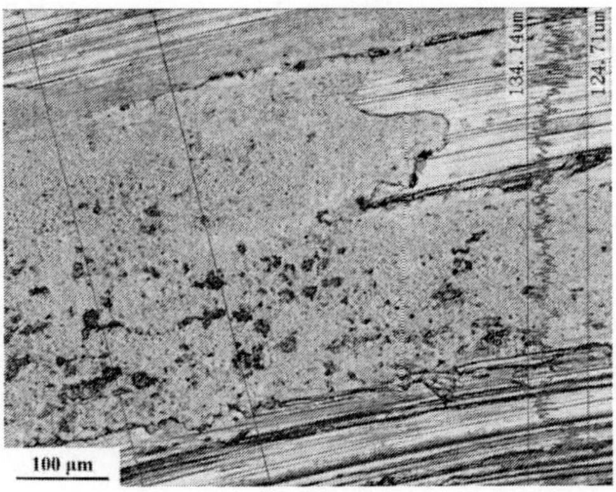

Figure 8.31: CLSM Image of the Surface Topography of the SLM Sample [18]

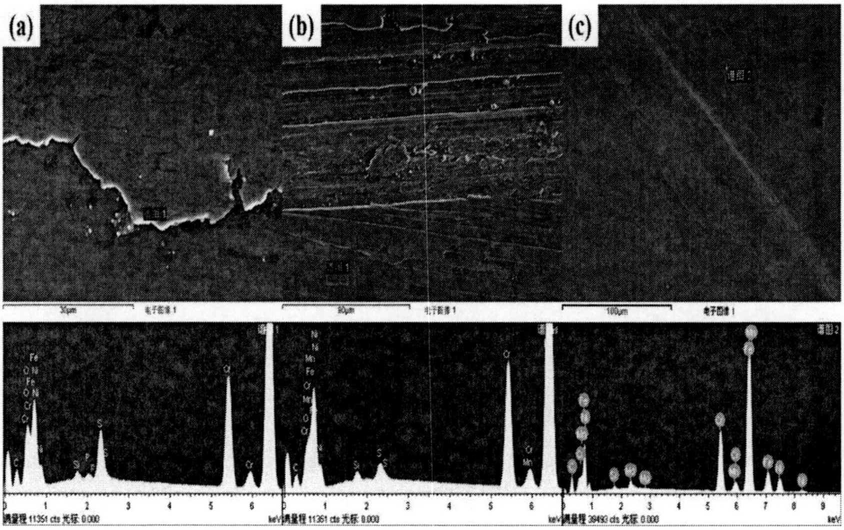

Figure 8.32: EDS/SEM Results [18]

8.6.2. *Corrosion*

Papula, et al [23] investigated the Duplex Stainless Steel 2205: Effect of Post-Processing Heat Treatment on Microstructure, Mechanical Properties, along with Corrosion Resistance processed by SLM.

Pre-alloyed, gas-atomized powder of SAF 2205 (UNS S31803/EN 1.4462) duplex stainless steel, purchased from Sandvik Osprey Ltd., Neath, UK, was used in this study with a particle size of 15 and 45 μm. 10 mm × 10 mm × 10 mm cubes are processed for corrosion test with a laser power 250 W, scan speed of 850 mm/s, layer thickness of 50μm, with a hatch pacing of 100 μm have been used as a optimized process parameter.

Corrosion resistance of the SLM-processed materials was studied by cyclic Potentiodynamic polarization (CPP) testing. Three different types of specimens were studied: SLM-processed as-built material, SLM-processed heat-treated (1000 °C/5 min) material, and a conventional cold-rolled and annealed duplex stainless steel 2205. Electrochemical testing was performed at room temperature using a potentiostat Interface 1010E (Gamry Instruments, Warminster, PA, USA). A three-electrode electrochemical cell containing 300 mL of neutral 0.6 M NaCl electrolyte was used with an exposed area of 1 cm^2. Duplex stainless steel specimens were the working electrode, saturated calomel electrode (SCE) as reference electrode, and P_t-wire as counter electrode, respectively. CPP curves were recorded at a scan rate of 1 mV/ s, after 1 h stabilization at open circuit potential (OCP). The scan range was set as follows: Start at 300 mV below OCP (cathodic potential) and the direction of polarization was reversed at 1.0 VSCE. Two replicates were performed for each sample to ensure the reproducibility. The tendency to pitting corrosion was estimated based on the characteristics of the CPP curves. Results of corrosion testing are shown in figure 8.33.

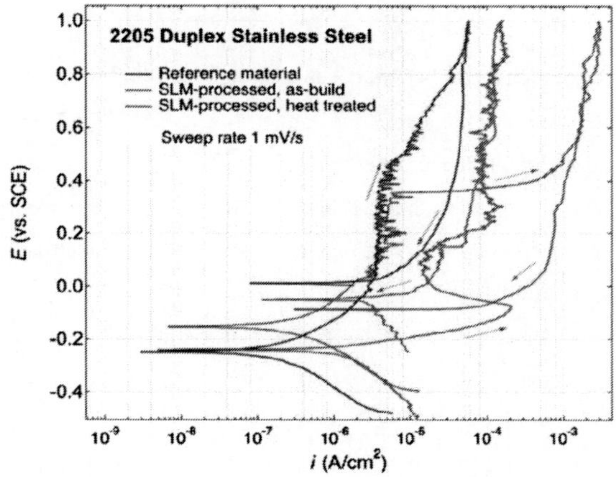

Figure 8.33: Cyclic Potentiodynamic Polarization Curves [23]

During the cyclic Potentiodynamic polarization in the potential range from 300 mV below OCP up to 1.0 VSCE, pitting breakdown potential (Epit) of 0.4 VSCE was observed for the SLM-processed as-built material. Marked hysteresis was observed when the direction of polarization was reversed and the return polarization curve followed an active path. The re-passivation potential of the SLM-processed as-built material was low, and the reduced re-passivation potential may be related to the porosity of the material. Pitting corrosion resistance of the SLM-processed material was significantly improved by the heat treatment, based on the pitting breakdown potential and hysteresis. After the heat treatment of the SLM processed material, similarly to the conventionally produced reference 2205 material, no distinct pitting initiation and hysteresis of the cyclic polarization curves in the studied potential range were observed, in contrary to the SLM as-built sample. Both of these samples have austenitic-ferritic microstructure, contrary to the as-built SLM sample. The conventionally produced DSS 2205 and the SLM-processed as-built material show a lower passive current density than the heat-treated SLM processed material. Reasons for this require further investigation.

The as-built SLM-processed material was almost fully (99%) ferritic, whereas after the heat treatments the austenite content increased to 40–46%. In general, duplex stainless steels exhibit greater corrosion resistance than single-phase austenitic and ferritic stainless steels in chloride-rich media [24]. This is predominantly due to the high alloy content of Cr and Mo. Recent research has indicated that the galvanic interaction between the two phases favors the enhancement of the passive film by modifications of its chemical composition and semi-conductive properties [25].

References

[1] https://www.imetllc.com/training-article/introduction-stainless-steel/

[2] https://www.asminternational.org/documents/10192/1849770/06940G_Chapter_1.pdf/53f29213-5dd6-4499-9959-841477b385b9

[3] Li, R., Shi, Y., Wang, Z., Wang, L., Liu, J., & Jiang, W, "Densification behaviour of gas and water atomized 316L stainless steel powder during selective laser melting", *Applied Surface Science,* 256(13), 2010, pg. 4350–4356.

[4] Gu, D., & Shen, Y, "Balling phenomena in direct laser sintering of stainless steel powder: Metallurgical mechanisms and control methods". *Materials & Design,* 30(8), pg. 2903–2910.

[5] Gu, D., & Shen, Y, "Balling phenomena during direct laser sintering of multi-component Cu-based metal powder." *Journal of Alloys and Compounds,* 432(1-2), 2007, pg. 163–166.

[6] Laohaprapanon, A., Jeamwatthanachai, P., Wongcumchang, M., Chantarapanich, N., Chantaweroad, S., Sitthiseripratip, K., & Wisutmethangoon, S, "Optimal Scanning Condition of Selective Laser Melting Processing with Stainless Steel 316L Powder", *Advanced Materials Research,* 341-342, 2011, pg. 816–820.

[7] K.A. Mumtaz, P. Erasenthiran and N. Hopkinson, "High density selective laser melting of Waspaloy", *J. Mater Process Tech.* 195, 2008, pg. 77-87.

[8] B. Vandenbroucke, *"Selective laser melting of biocompatible metals for rapid manufacturing of medical parts",* K.U. Leuven. Facultenit Ingenieurswetenschppen publisher, 2008, pg. 43.

[9] D. Gu and Y. Shen, "Balling phenomena in direct laser sintering of stainless steel powder: Metallurgical mechanisms and control methods", *J. Mater design.* 30, 2009, pg. 2903-2910.

[10] Shang, Y., Yuan, Y., Li, D., Li, Y., & Chen, J, "Effects of scanning speed on in vitro biocompatibility of 316L stainless steel parts elaborated by selective laser melting", *The International Journal of Advanced Manufacturing Technology,* 92(9-12), 2017, pg. 4379–4385.

[11] Ciocan LT, Miculescu F, Patrascu I, Miculescu, M. *"Biological reactions to dental implants",* 2011, INTECH Open Access Publisher

[12] Geenen, K., Röttger, A., Feld, F., & Theisen, W, "Microstructure, mechanical, and tribological properties of M3:2 high-speed steel processed by selective laser melting, hot-isostatic pressing, and casting", *Additive Manufacturing* 28, 2019, pg. 585–599.

[13] H.P. Tang, M. Qian, N. Liu, X.Z. Zhang, G.Y. Yang, J. Wang, Effect of powder reuse times on additive manufacturing of Ti-6Al-4V by selective Electron beam melting, the journal of the minerals, *Metals Mater. Soc.* 67, 2015, pg. 555–563.

[14] L.C. Ardila, F. Garciandia, J.B. González-Díaz, P. Álvarez, A. Echeverria, M.M. Petite, R. Deffley, J. Ochoa, Effect of IN718 recycled powder reuse on properties of parts manufactured by means of selective laser melting, physics procedia, 8th *International Conference on Laser Assisted Net Shape Engineering (LANE 2014)* vol. 56, 2014, pg. 99–107.

[15] A. Röttger, K. Geenen, M. Windmann, F. Binner, W. Theisen, "Comparison of microstructure and mechanical properties of 316L austenitic steel processed by selective laser melting with hot-isostatic pressed and cast material", *Mater. Sci. Eng.* A 678, 2016, pg. 365–376.

[16] J. Krell, A. Röttger, K. Geenen, W. Theisen, General investigations on processing tool steel X40CrMoV5-1 with selective laser melting, *J. Mater. Process. Technol.* 255, 2018, pg. 679–688.

[17] Alsalla, H.H., Smith, C., & Hao, L, "Effect of build orientation on the surface quality, microstructure and mechanical properties of selective laser melting 316L stainless steel", *Rapid Prototyping Journal,* 24(1), 2018, pg. 9–17.

[18] Zhu, Y., Zou, J., Chen, X., & Yang, H, "Tribology of selective laser melting processed parts: Stainless steel 316L under lubricated conditions", Wear, 350-351, 2016, pg. 46–55.

[19] B. Wang, *"Lubrication characteristics and parameter optimization of barrel-port plate in an axial piston pump",* Zhejiang University, China, 2009.

[20] Y. Zhu, X. Chen, J. Zou, H. Yang, A study on the influence of surface topography on the low-speed tribological performance of port plates in axial piston pumps, *Wear.* 338-339, 2015, pg. 406–417.

[21] A. Beek, *"Advanced engineering design",* 2006th ed., TU Delft, the Netherlands, 2006.

[22] Y. Zhu, X. Chen, W. Wang, H. Yang, A study on iron oxides and surface roughness in dry and wet wheel–rail contacts, *Wear.* 328-329, 2015, pg. 241–248.

[23] Papula, Song, Pateras, Chen, Brandt, Easton, Hänninen, "Selective Laser Melting of Duplex Stainless Steel 2205: Effect of Post-Processing Heat Treatment on Microstructure, Mechanical Properties, and Corrosion Resistance", *Materials,* 12(15), 2019, 2468.

[24] Moura, V.S.; Lima, L.D.; Pardal, J.M.; Kina, A.Y.; Corte, R.R.A.; *"Tavares, S.S.M. Influence of Microstructure on the Corrosion Resistance of the Duplex Stainless Steel UNS S31803".* Mater. Charact, 59, 2008, pg. 1127–1132.

[25] Cheng, X.; Wang, Y.; Li, X.; Dong, C, "Interaction between Austenite-Ferrite Phases on Passive Performance of 2205 Duplex Stainless Steel". *J. Mater. Sci. Technol,* 34, 2018, pg. 2140–2148.

Chapter IX

Selective Laser Melting (SLM) of Composites

Chapter IX

9. Selective Laser Melting (SLM) of Composites

9.1. Introduction

A composite is nothing but a combination of two or more materials, commonly referred to as constituents, and have material properties derived from the individual constituents. These properties may have the combined characteristics of the constituents or they are substantially different. Sometimes the material properties of a composite material may exceed those of the constituents.

The properties of the composite material and those man-made composites are produced by combining the constituents by various means with respect to the constituents have to be present in reasonable proportions that the constituent phases with quite different properties as shown in figure 9.1 with (a) Laminate with uni directional or bidirectional layers, (b) irregular reinforcement with long fibres, (c) reinforcement with particles, (d) reinforcement with plate strapped particles, (e) random arrangement of continuous fibres, (f) irregular reinforcement with short fibres, (g) spatial reinforcement, (h) reinforcement with surface tissues as mats, woven fabrics, etc.

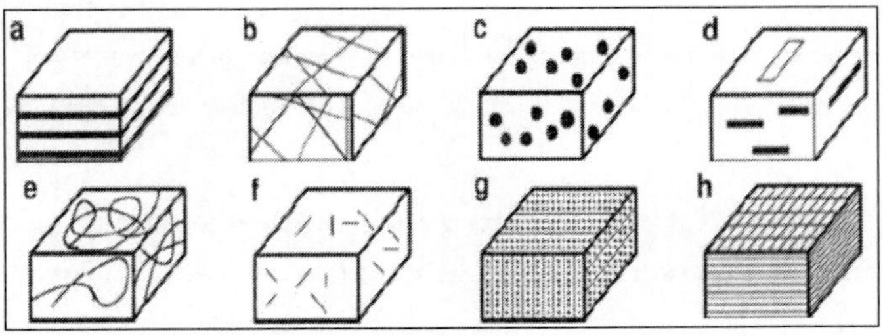

Figure 9.1: Examples of Composite Materials with different forms of Constituents and Distributions of the Reinforcements [1]

It consists of mainly of two materials of different nature, one material is discontinuous and is called the reinforcement, and the other material is mostly less stiffer and weaker. It is continuous and is called the matrix. The properties of a composite material depends on

- The properties of the constituents.
- The geometry of the reinforcements, their distribution, orientation and concentration usually measured by the volume fraction or fiber volume ratio.
- The nature and quality of the matrix-reinforcement interface.

Composites can be classified by their arrangement form and the distribution of their constituents as shown in figure 9.2.

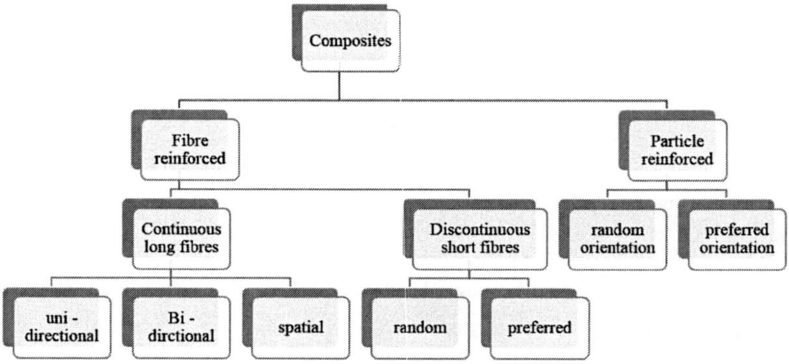

Figure 9.2: Classification of Composites based on their Arrangements [1]

The reinforcement constituent can be described as fibrous or particulate. The fibres are continuous (long fibres) or discontinuous (short fibres). Long fibres are arranged usually in uni or bidirectional, but also irregular reinforcements by long fibres are possible. The arrangement and the orientation of long or short fibres determine the mechanical properties of composites and the behaviour ranges between a general anisotropy to a quasi-isotropy. Particulate reinforcements have different shapes, they may be spherical, platelet or of any regular or irregular geometry. Their arrangement may be random or regular with preferred orientations [2, 3].

The metals initially used for MMC matrix materials generally were traditional alloys. As time has progressed, however, special matrix materials, tailored for use in composites, have been developed. The key metallic matrix materials used for structural MMCs are alloys of aluminium, titanium, and iron. There was a significant amount of research on composites using intermetallic compound matrix materials, such as titanium aluminides, but these were largely unsuccessful. There have been many other metals used as matrix materials, including copper, lead, magnesium, cobalt, silver, and super alloys. The in situ properties of metals in a composite depend on the manufacturing process and, because metals are elastic–plastic materials, the history of mechanical stresses and temperature changes to which they are subjected.

One of the major advantages of metal matrix composites reinforced with continuous fibres over unidirectional polymer matrix composites is that many, if not most, of the former have much greater transverse strengths, which allows them to be used in a unidirectional configuration.

9.2. Mechanical Properties of Composites

Composites are multifunctional materials having unprecedented mechanical and physical properties which can be tailored to meet the requirements of a particular application. Many composites also exhibit great resistance to wear, corrosion, and high-temperature exposure. [4]

9.2.1. Particle Reinforcement

In general, particle reinforcement is strengthening the composites less than fiber reinforcement. It is used to enhance the stiffness of the composites while increasing the strength and the toughness. Because of their mechanical properties, they are used in applications in which wear resistance is required. For example, hardness of cement can be increased by reinforcing gravel particles, drastically. Particle reinforcement a highly advantageous method of tuning mechanical properties of materials since it is very easy implement while being low cost. [5]

9.2.2. Continuous Fiber Reinforcement

Continuous fiber reinforcement is implemented by incorporating a fiber as the strong phase into a weak phase, matrix. The reason for the popularity of fiber usage is materials with extraordinary strength can be obtained in their fiber form. Non-metallic fibers are usually showing a very high strength to density ratio compared to metal fibers because of the covalent nature of their bonds. The most famous example of this is carbon fibers that have many applications extending from sports gear to protective equipment to space industries [6]. The stress on the composite can be expressed in terms of the volume fraction of the fiber and the matrix.

9.2.3. The Effect of Fiber Orientation

The change is in the fiber orientation can affect the mechanical properties of the fiber-reinforced composites especially the tensile strength. The composite tensile strength can be predicted to depending on the θ (0° to 10° angles), the angle between the applied for, and the orientation of the fibers.

9.2.4. Mechanical Properties various Materials

Table 9.1 shows the one of the most important features and advantage of fiber composites over metal, that is specific strength and specific stiffness. Although the steel and the aluminium alloy have comparable strength and stiffness with fiber composites, the specific strength and stiffness of composites are around higher than steel and the aluminium.

Table 9.1: Comparison of Cost, Specific Strength, and Specific Stiffness [7]

	Carbon Fiber Composite (aerospace grade)	Carbon Fiber Composite (commercial grade)	Fiberglass Composite	Aluminium 6061 T-6	Steel, Mild
Cost $/LB	$20 - $250+	$5 - $20	$1.50 - $3.00	$3	$0.30
Strength (psi)	90,000 - 200,000	50,000 - 90,000	20,000 - 35,000	35,000	60,000
Stiffness (psi)	10×10^6- 50×10^6	$8 \times 10^6 - 10 \times 10^6$	1×10^6 - 1.5×10^6	10×10^6	30×10^6
Density (lb/in³)	0.050	0.050	0.055	0.10	0.30
Specific Strength	$1.8 \times 10^6 - 4 \times 10^6$	$1 \times 10^6 - 1.8 \times 10^6$	363,640–636,360	350,000	200,000
Specific Stiffness	200×10^6 - $1{,}000 \times 10^6$	160×10^6- 200×10^6	18×10^6- 27×10^6	100×10^6	100×10^6

9.3. Applications of Composites

Composites are versatile, used for both structural applications and components, in all aircraft and spacecraft, from hot air balloon gondolas and gliders, to passenger airliners, fighter planes and the Space Shuttle.

9.3.1. Aerospace

Weight is everything when it comes to heavier-than-air machines, and designers have striven continuously to improve lift to weight ratios since man first took to the air. Composite materials have played a major part in weight reduction, and today there are three main types in use: carbon fiber-, glass- and aramid- reinforced epoxy; there are others, such as boron-reinforced (itself a composite formed on a tungsten core). Since 1987, the use of composites in aerospace has doubled every five years, and new composites regularly appear.[8]

9.3.2. Automotive/ Transportation/Farm/Construction/Infrastructure

Composites are now being used in vehicle and equipment applications, including, panels, frames, interior components and other parts [9, 10]. Some composite infrastructure applications include buildings, roads, bridges and pilings. Thermoset composites are replacing many traditional materials for home and offices architectural components including fixtures, doors, wall panels, roofing, window frames, moulding, vanity sinks, shower stalls and even swimming pools. [11]

9.3.3. *Corrosive Environments in Chemical Industries*

Composites are ideal for applications in corrosive environments, such as chemical processing plants, pulp and paper converting, oil and gas refineries and water treatment facilities. Common applications include fans, grating, tanks, ducts, hoods, pumps and cabinets. [12].

9.3.4. *Marine*

With their corrosion resistance and light-weighting attributes, Marine composite applications include boat hulls, bulkheads and other components for military, commercial and recreational boats and ships [13].

9.4. Densification Behaviour based on Laser Related Process Factors

Gu, D et al [14] investigated in-situ of TiC/Ti5Si3 composites with novel reinforcement architecture and elevated performance using Selective Laser Melting with 99.9% purity, poly-angular Ti powder with an average particle size of 45 μm and 99.0% purity, irregular-shaped SiC powder with a mean particle size of 13 μm. The original Ti and SiC ingredients were mixed according to a weight ratio of 76.2:23.8 (an equivalent molar ration of 8:3) using high energy ball mill for 15 hours at 300 rpm speed with ball to powder weight ratio of 10:1.

The SLM process is carried using YLR – 200μm Ytterbium fiber laser with a power of 200W used fixed power of 80W, spot size of 70μm, layer thickness of 50μm and inert gas atmosphere (30 bar pressure) is maintained with a oxygen content of <10ppm.The scan rate were 0.1,0.2, 0.3 and 0.4m/s during a batch of experiment and the energy densities (η) of 0.2,0.267,0.4 and 0.8kJ/m were used and the specimens of 8mm×8mm×8mm was prepared.

Figure 9.3 depicts the influence of laser energy density on densification of SLM-processed parts. On increasing the applied η from 0.2 to0.267 kJ/m, the densification level increased from 91.6% to 95.7%, which was further elevated to a maximum value of 96.9% at a higher η of 0.4 kJ/m. However, as a considerably high η of 0.8 kJ/m was used, the densification rate decreased sharply to 87.2%. The corresponding cross-sectional microstructures showed that a number of small-sized interlayer pores with an average size of ~20 μm were formed at a relatively low η of 0.2 kJ/m. The considerably dense SLM layers without any apparent pores or cracks were produced as the applied η increased ≥0.267 kJ/m. A significant decrease in the obtained density at 0.8 kJ/m was due to the formation of interconnected cracks vertical to SLM layers, with the length in the order of millimetre. The success of SLM densification is determined by the wettability of the primarily precipitated TiC phase by the residual Ti–Si liquid in SLM system.

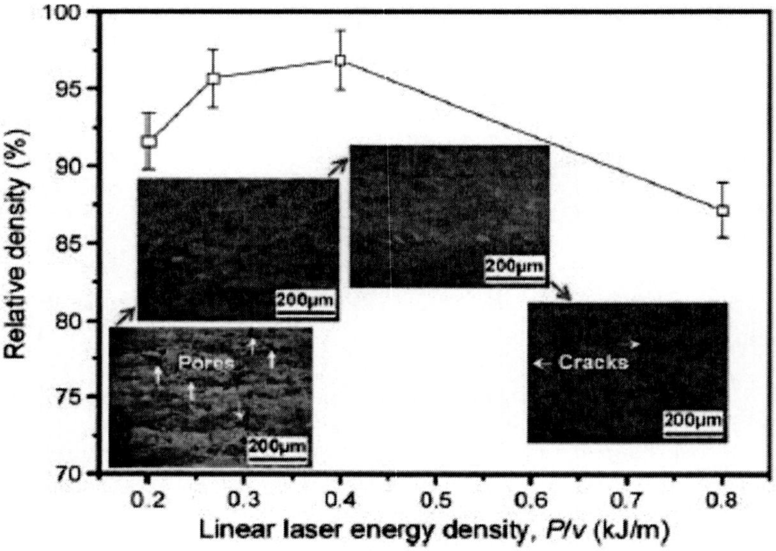

Figure 9.3: Influence of Laser Energy Density on the Densification Rate and Cross Sectional Microstructure [14]

According to Hosson et al.'s results [15], for a reactive wetting system of liquid metals on ceramic phase, the wettability is predominated by the formation of ceramic phase during processing. The equilibrium contact angle θeq can be approximately written as:

$$\text{Cos}\ \theta_{eq} = (\gamma_{sv} - \gamma_{1s}) / \gamma_{1v} \qquad (9.1)$$

Where, γ_{lv}, γ_{sv}, and γ_{ls} are the interface tensions of liquid metal-vapour, reacted ceramic-vapour, and liquid metal-reacted ceramic, respectively. γ_{lv} may vary during reaction as the reaction product may diffuse into the liquid [15].

For Ti–Si liquid system, γ_{ls} is dependent with the operating temperature T [16]:

$$\gamma_{1s} = 1656 - 0.375T \qquad (9.2)$$

An increase in the applied η to 0.4 kJ/m tends to elevate the workable T, which in turn lowers γls. Therefore, a lower contact angle θ in equation 9.1 and, accordingly, an improved liquid–solid wettability is achieved, favouring a sufficient consolidation of the composite system during SLM. On the other hand, the shrinkage rate $d(\Delta L/L_0)/dt$ during solidification can be estimated by [17]:

$$\frac{d(\frac{\Delta L)}{L0}}{dt} = \frac{\Delta PW}{2R\mu} \quad (9.3)$$

Where, ΔP is the capillary pressure, R the grain radius, W the liquid thickness, and μ is the liquid viscosity, μ can be with respect to the research paper [18] and it is defined by:

$$\mu = \frac{16}{15}\sqrt{\frac{m}{kT}}\gamma_{ls} \quad (9.4)$$

where m is the atomic mass and k is the Boltzmann constant. As a considerably high η of 0.8 kJ/m is applied for SLM, the resultant higher T and lower γ_{ls} both result in a decrease in μ (Eq. (9.4)), hence increasing the shrinkage rate during solidification (Eq. (9.3)). A large amount of thermal stresses tend to generate in the solidified parts [19], resulting in the formation of thermal cracks and the decrease in densification.

9.5. Defects in SLM of Composites

9.5.1. Porosity

Table 9.2: Chemical Composition (in weight %) of the Standard Alloys [20]

Element	Si	Fe	Cu	Mn	Mg	Other	Al
Al	-	-	-	-	-	-	99.7 purity
AlMg1SiCu	0.40- 0.80	≤0.70	0.150.40*	0.15	0.80-1.20	≤0.6	Balance
AlSi10Mg	9.0-11.0	≤0.55	≤0.05	≤0.45	0.2-0.45	≤0.40	balance

Dadbakhsh, et al [20] investigated the effect of Al alloys on selective laser melting behaviour and microstructure of in situ formed particle reinforced composites with a powder mixtures of pure Al (mean particle size ~40 μm - Alpoco, UK), AlMg1-SiCu (mean particle size ~40 μm - Alpoco, UK), and AlSi10Mg (mean particle size~40 μm - EOS GmbH, Germany) with 15 wt% Fe_2O_3 powder (sieved below 53 μm- Inoxia, UK) were prepared. The mixing was carried out by blending and vibrating the mixture at least 30 min using a test sieve shaker. The homogenous mixing appeared in a uniform reddish colour of the powder mixtures. The desegregation was not an issue in blending since tiny pieces of Fe2O3 powder tended to adhere to Al powder surface. The chemical compositions of these Al alloys are shown in table 9.2, according to the used standards by powder manufacturers.

An SLM machine (Realizer 250, MCP Ltd.) was used to process powder mixtures. Argon gas was pumped into the build chamber to keep the O_2 level below 0.9%. Multilayer SLM samples were fabricated using spot size of 0.16 mm, scan line spacing of 0.05 mm, layer thickness of 0.075 mm, laser power of 74 W, and scanning speed of 0.20 m/s (parameters were picked on the basis of the previous experiences to reach a good hardness and composite microstructure). Twice scanning was carried out for every layer. The scanning strategy was alternating [21], i.e., x direction for the first layer and subsequently y direction for the next layer, and so on.

After SLM of different powder mixtures, the influence of Al alloy on porosity formation was studied using CT scan images. Figure 9.4 (a) shows broad spread of porosity in three dimensions of the SLM parts made from Al/15wt%Fe_2O_3. The extensive porosity can be attributed to the formation of oxides and their displacement into inter-solid regions. Louvis state that if these oxides do not vaporise or stir into the melt pool, the melt pool fails to fuse and wet surrounding regions, leading to a porous structure.

In other words, the accumulated oxides act as barriers suppressing diffusion and melting fusion (which joins the solids together) and thus leaving much porosity within the part. The change of the Al alloy can manipulate the porosity formation. The usage of AlMg1- SiCu instead of pure Al mixed with 15 wt% Fe_2O_3 seems to slightly mitigate the porosity formation (figure 9.4b). This reduction in porosity becomes even more evident when the AlSi10Mg/15wt%Fe_2O_3 is used (figure 9.4 b & c).

This reduction in inter-solid porosity with changing the Al alloy can be attributed to improvement of melting due to changing the thermal behaviour in presence of higher content of alloying elements. For example, AlSi10Mg has a lower thermal conductivity, reducing the cooling rate and leading to higher melting pool temperature. This will decrease the surface tension and increase the convection flow which intensifies tearing the oxide barriers (accumulated in inter-solid regions) apart and improves bonding between solids.

In addition, Si promotes the breakdown of the protective oxide layers in Al– Mg–Si alloys, which contributes to better bonding. As observed from CT scanning results, the largest difference in porosity formation was between SLM parts made from Al/ 15wt%Fe_2O_3 and AlSi10Mg/15wt%Fe_2O_3.

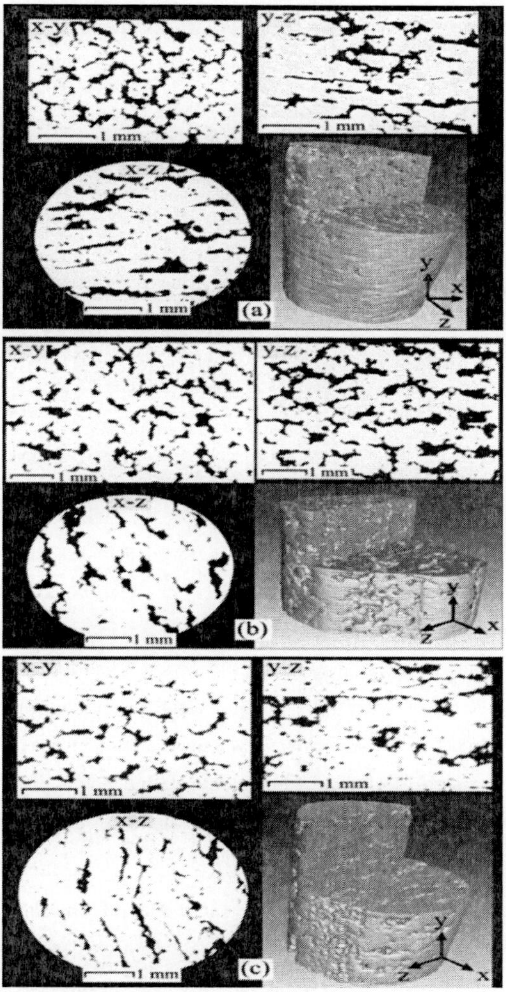

Figure 9.4: CT Scanning Exhibition of the SLM Parts [21]

This difference is better shown in figure 9.5, where the closer demonstration of their x-y cross section was exhibited. As seen, the inter-solid porosity in Al/15wt%Fe_2O_3 (figure 9.5a) changes to more spherical pores in AlSi10Mg/15wt%Fe_2O_3 (figure 9.5b). As mentioned above, this can be attributed to changing in the thermal behaviour and stability of oxide layer with additional alloying elements. However, very tiny micro pores can also form in microstructure as well. These tiny pores (figure 9.5 b) seem to originate from gas; when excessive hydrogen is rejected from solution and it recombines as molecular gas, it can be entrapped in the solid structure and forms tiny pores.

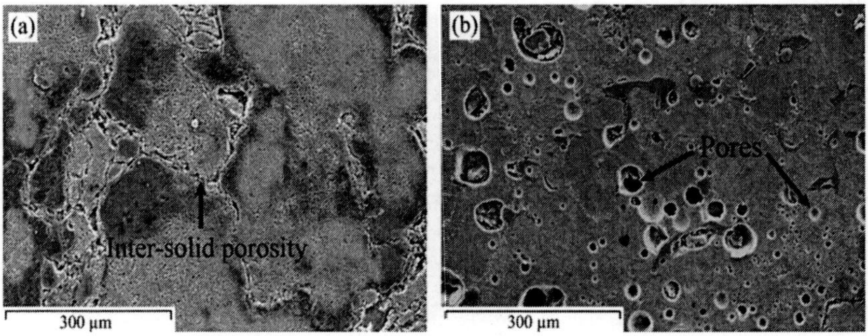

Figure 9.5: SEM View of the x-y Cross Section of the SLM Parts [21]

9.5.2. Crack

Li, X. P., et al [22] investigated the role of a low-energy–density re-scan in fabricating crack-free $Al_{85}Ni_5Y_6Co_2Fe_2$ bulk metallic glass composites via selective laser melting with the $Al_{85}Ni_5Y_6Co_2Fe_2$ (in at.%) metallic glass (MG) powder used in this work was nitrogen-atomized after arc melting high-purity elemental pieces Al (>99.9 wt.%), Ni (>99.9 wt.%), Y (>99.0 wt.%), Co (>99.9 wt.%) and Fe (>99.0 wt.%) under a Ti-gettered argon atmosphere for six times to ensure chemical homogeneity. Then the powder was sieved and the particles below 45 lm were used. As a result of the relatively low cooling rate within powder particles during gas-atomisation, the powder is partially crystallized and contains face centred cubic (fcc) Al crystals and other intermetallic [23].

SLM experiments were conducted using a ReaLizer SLM-100 machine (ReaLizer GmbH, Germany) which is equipped with a fibre laser, generating a laser beam with a wavelength of 1060 nm and maximum power of 200 W at the part bed. An inert high purity argon gas atmosphere was used during the whole experiment to prevent oxidation. The laser scan speed and power for the initial scan were 625 mm/s and 200 W, respectively. A laser power of 80 W was used for the re-scan at the same scanning speed. The powder layer thickness was fixed at 50 lm, the scan spacing was 150 μm and substrate was unheated. Small cubes with dimensions 7 ×7× 5 (height) mm were prepared.

The amorphous nature and phase formation of the fabricated BMGs were examined and characterised using X-ray diffraction (XRD, PANalytical Empyrean, D/max III, Cu Ka target, operated at 40 kV and 40 mA with a step size of 0.02° and scanning speed 2°/min). Scanning electron microscopy (SEM, Zeiss 1555 VP-FESEM, operated at an accelerating voltage of 15 kV and working distance 10 mm) was used to investigate the formation and growth of the cracks and the microstructure evolution of the BMGCs.

Figure 9.6 illustrates the typical resultant thermal stress within each layer of an SLM-processed component [19]. Due to the rapid solidification arising from the high cooling rate and the constraint from neighbouring solidified material, the currently-scanned powder layer will have tensile thermal stress at the bottom and a compressive stress at the top.

As the next powder layer is scanned, this thermal stress undergoes a series of partial relief and building-up process and accumulates layer after layer. This thermal stress is widely-known to cause the micro- and macro-cracks within components fabricated using SLM.

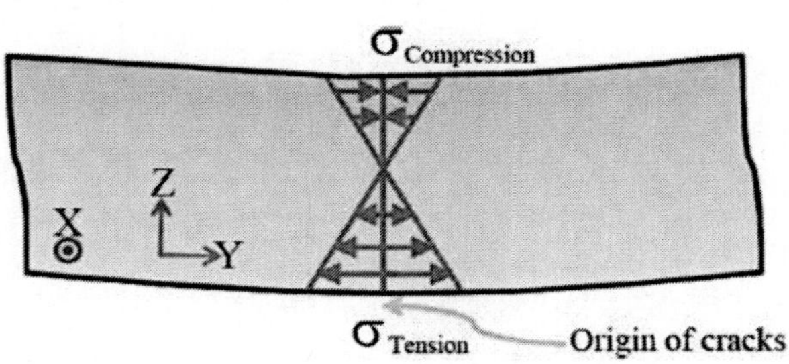

Figure 9.6: Schematic Illustration of the Thermal Stress and the Origin of Cracks within each Layer of an SLM-processed Component [22]

A photo of an $Al85Ni5Y6Co2Fe_2$ BMGC, built using a conventional single scan strategy, is shown in figure 9.7a. It can be clearly seen that a macro-crack has formed at the very bottom and extends through the middle of the sample. The crack has started from the first solid layer of the component rather than in the support and propagates along the building direction (Z-axis), as shown by the crack tip in figure. 9.7b. This indicates that, unlike ductile materials where the support can be used to help withstand the thermal stresses and therefore prevent macro-cracks, the same appears not to be true in BMGCs. In this work, the first layer was built by laser scanning in X-axis direction, where the maximum thermal stress is expected to be perpendicular to the X-axis [24]. Consequently, the macrocrack forms perpendicular to the X-axis (figure 9.7a). When the thermal stress exceeds the yield stress of ductile materials, the stress can be relieved through plastic deformation.

However, if the thermal stress increases further such that it exceeds the tensile strength, cracks will initiate and propagate until the strength of the layers can withstand the accumulated thermal stress. For intrinsically brittle BMGCs, the thermal stress will cause the formation and propagation of shear bands and when the stress exceeds the yield strength, cracking and catastrophic failure will occur (figure 9.7c) [25]. In order to produce crack-free material, it is therefore necessary to either increase the strength of the material or decrease the residual stress. The former is not easy and is largely a function of the material. Reducing the residual stress may be possible by reducing the amount of incident energy, however this is likely to be at the detriment of the density (and therefore strength) of the part.

Another option may be to undertake in-situ stress relieving within each layer. To test this, a dual scanning strategy was used which incorporated a normal high energy scan to produce high density, followed immediately by re-scanning using a lower laser energy. The result of this on an $Al_{85}Ni_5Y_6Co_2Fe_2$ BMGCs is shown in figure 9.7d. The first (lower) half of the sample (below the arrow in figure 9.7d) was scanned using the normal single scan strategy, after which a second, low energy scan was added to each layer. It can be clearly seen that the crack tip stops propagating at the layer where this re-scanning is first carried out, as shown in figure 9.7e and f. Conclusion has been made that the influence of a low laser-energy–density rescan on stopping macro-crack propagation in $Al_{85}Ni_5Y_6Co_2Fe_2$ BMGCs during SLM process.

The underlying reason for this technique stopping the cracking has been ascribed to: (1) the formation of super plasticity during re-scanning where the initial residual stress is rapidly relieved through viscous flow; and (2) reduction of thermal stress after re-scanning due to lowered temperature gradient (and therefore cooling rate). These both result in a relief and reduction of thermal stress which accumulates during SLM, facilitating the fabrication of intrinsically brittle BMGCs free from macro-cracks.

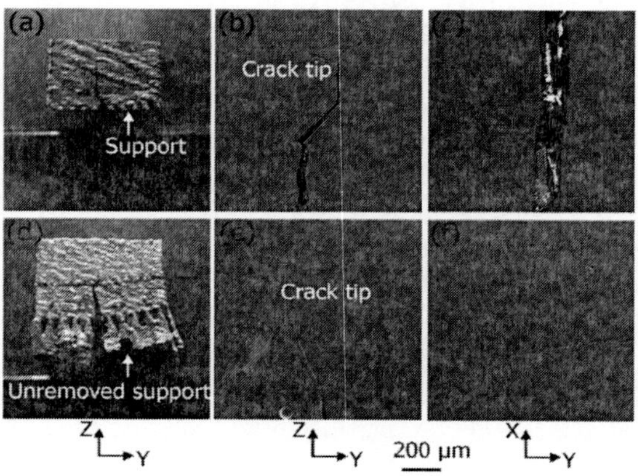

Figure 9.7: Specimen Built with and without Rescanning Process [22]

9.6. Mechanical Properties of SLMed Part or Component

9.6.1. *Hardness*

Gu, D et al [14] investigated hardness of SLM processed $TiC/TiSi_3$ along with the effect of energy densities which was discussed in section 9.4 The average microhardness measured on the cross-sections of SLM-processed $TiC/Ti5Si_3$ composite parts, as shown in figure 9.8, exhibited a similar trend as the relative density with respect to the laser energy density. A maximum micro hardness value of 980.3HV0.2 was obtained at 0.4 kJ/m. In this situation, a small-sized indentation with a clear configuration was observed, indicating a small degree of deformation during indentation and an elevated hardness performance.

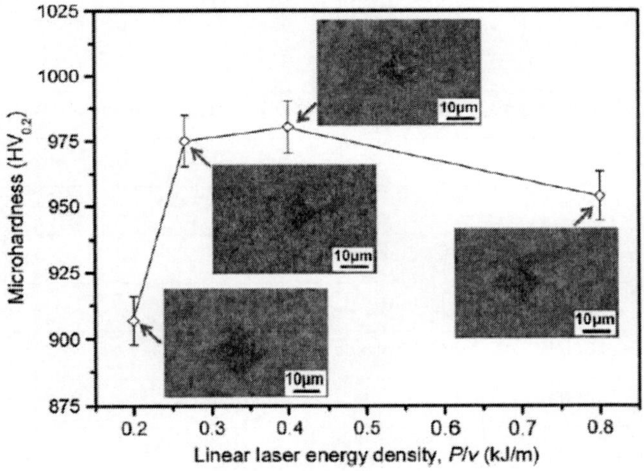

Figure 9.8: Effect of Laser Energy Density on Micro Hardness and Indentation Morphology of SLMed Processed Part [14]

For comparison, the pure Ti parts without the addition of SiC were prepared under the same SLM processing conditions and the average micro hardness was 281.0HV0.2. The micro hardness of in-situ $TiC/Ti5Si_3$ composites prepared by SLM of SiC /Ti system showed more than a 3-fold increase upon that of the unreinforced Ti part.

Attar, H., et al [26] investigated the Selective laser melting of in situ titanium–titanium boride composites, Processing, and mechanical properties. Commercially pure Ti (CP-Ti) powder (99.7% purity, d50 = 48.69 lm) and TiB2 powder (98.9% purity, d50 = 3.5–6 lm) were used in this study. The powders were mixed under protective argon atmosphere to obtain a Ti– 5 wt.%

TiB2 mixture and poured into ball-mill vials. C15 carbon steel balls with a diameter of 10 mm were used and the ball-to-powder weight ratio was set to 5:1. Afterwards, the powder mixture was ball-milled using a Retsch planetary ball mill for 1–4 h at a fixed rotation speed of 200 rpm. A MTT SLM 250 HL machine containing a 400 W Yb:YAG fiber laser with an 80 lm spot size was used to produce several cylindrical parts under protective high-purity argon atmosphere. The samples were manufactured with various laser powers of 165–185 W, laser scanning speeds of 118–154 mm/s and fixed layer thickness and constant hatching distance of 100 μm. The layers were scanned using a continuous laser mode according to a zigzag pattern, which was alternated 90° between each layer. Vickers micro hardness testing was performed on polished samples using a HMV Shimadzu micro hardness tester with 50 g load and 10 s dwelling time. An average of 10 indents was taken for each measurements shown in table 9.3.

Table 9.3: Comparison of Vickers Hardness of Ti Materials [26]

Material type	**Processing method**	**Vickers hardness (Hv)**
Ti-8.35 vol.%TiB	SLM	402±7
CP- Ti	SLM	261±13
CP-Ti	Casting	210
Ti-25Pd-5Cr	Casting	261
Ti-6Al-4V	Superplastic forming	346
Ti-24Nb-4Zr-8Sn	SLM	227±8

As can be seen, the mean Vickers micro hardness of the Ti–TiB composite is 402 Hv, thus being larger than the corresponding values of as-cast and SLM-processed CP-Ti. Also, the Ti–TiB composite shows a higher micro hardness compared to conventional Ti implant alloys. The increase in micro hardness of the Ti– TiB composite compared with CP-Ti can be attributed to the hardening effect induced by the TiB particles and the refinement of Ti grains. It has been reported that the enhancement of hardness may improve the wear resistance of implants [27]. An improvement of wear resistance for Ti with addition of TiB was reported in Ref. [28]. The resulting reduction of wear debris may prevent loosening of implants and increase their longevity in the human body [29].

9.6.2. *Tensile Strength*

Song, B et al [30] investigated tensile behaviour of hybrid Nano-micro SiC reinforced iron matrix composites produced by selective laser melting with Fe powder (MH 300, Hoganas

Company), with an average particle size (D50) of 20 μm, and nano/micro-scaled SiC powder (Chaoxin Company, China) with an average particle size (D50) of about 0.7 μm, were used as feedstock. The characteristics of those two powders are shown in figure 9.9 which has (a) SEM image and XRD pattern of Fe powder, (b) SEM image and XRD pattern of SiC, (c) powder size distribution of the iron powder, and (d) powder size distribution of the nano/micro-scaled SiC powder. As shown by XRD patterns, a-Fe and b-SiC appear to be the only phase respectively for Fe and SiC powders. Before selective laser melting, Fe and SiC powders with the volume ratio of 95–5 (i.e., including 2.2 wt.% SiC) were blended in a tumbling mixer for 2 h.

A stainless steel plate with dimensions 300 mm × 300 mm×5 mm was used as the building platform. Before being installed, this platform was grit-blasted with alumina. A commercial SLM machine (Realizer 250, MCP Ltd.) was used. The laser source is YLR-100-SM single Mode CW Ytterbium fiber laser (1064–1100 nm). The working chamber provides a closed environment filled with argon with an outlet pressure of 10 mbar.

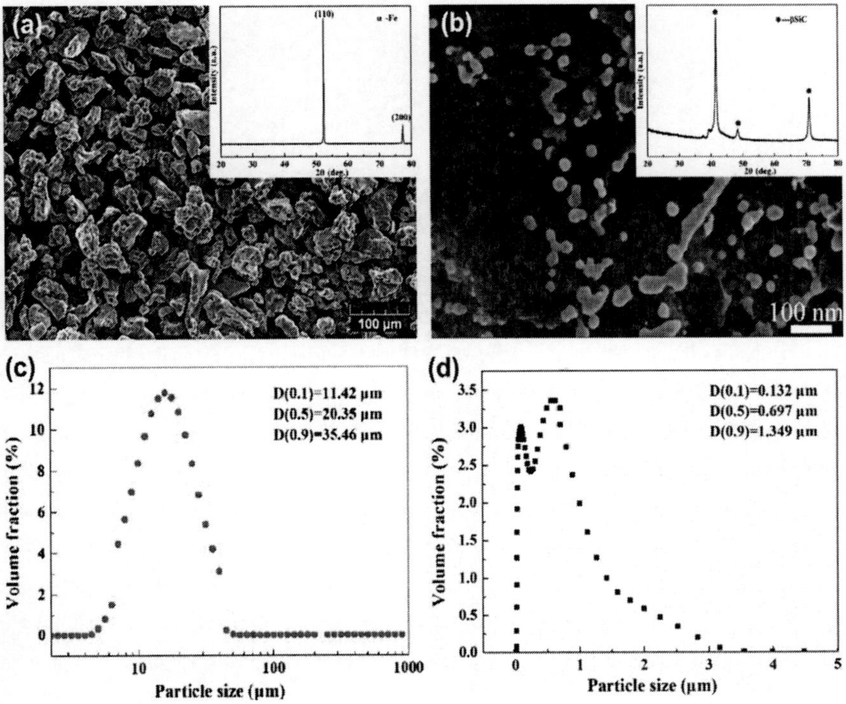

Figure 9.9: Characteristics of the Feedstock Powders [30]

The protective gas allows maintaining the oxygen content below 0.2% to prevent the oxidation of Fe during the fabrication process. On the basis of a series of preliminary experiments, SLM was performed to fabricate the cubes and the tensile specimens (figure 9.8) using the following parameters: diameter of the laser beam 34 μm, laser power 100 W, scan line spacing 40 μm, thickness of each powder layer 50 μm, scanning speeds of 0.33 m/s and zigzag pattern with 90° rotation between the adjacent layers. Tensile tests were performed on the SLM-fabricated composite specimen and the results were compared with those of as-fabricated pure iron sample produced under the same SLM conditions. Compared with the as-fabricated pure iron sample, three obvious deformation stages could be observed for the as-fabricated composite specimen as shown in figure 9.11. The first stage is a classical elastic stage going up to 120±10 MPa, which shows little of any difference between the composite material and pure iron.

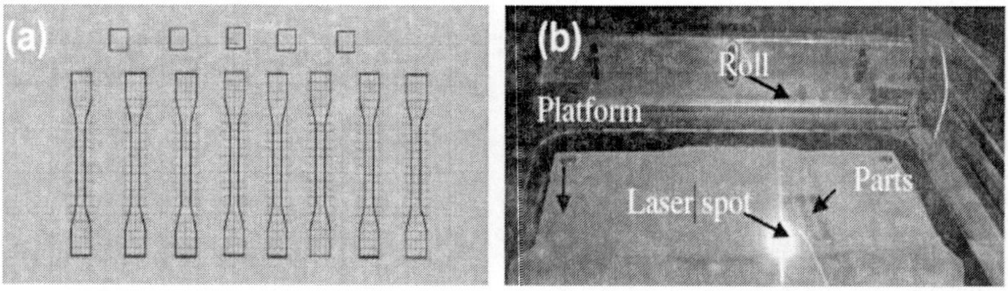

Figure 9.10: Images of Three-dimensional Model of Samples [30]

The elastic modulus of the as-fabricated specimens (pure iron or composite) is about of 210 ± 11 GPa. The second deformation stage for the composite is very different with that of pure iron. While the ultimate tensile strength (UTS) of the as-fabricated pure iron hardly reaches 357 ± 22 MPa after 9% strain, that of the composite material increases sharply from 357 ± 22 MPa to over 600 ±13 MPa with 5% strain. This hardening process is probably related to the dislocation pinning by the SiC particles. In the third stage, when the as-fabricated pure iron specimen breaks, strain hardening continues for the as-fabricated composite, although at a reduced rate vs. the previous stage, finally UTS reaches to 764 ±15 MPa. The existence of the last two stages could be attributed to the bimodal grain size of the reinforcing phase but at this point, this phenomenon could not be elucidated. The reasons for the observed improvement in yield strength and ultimate tensile strength for the as-fabricated composite are described in great detail in the literature [31-33].

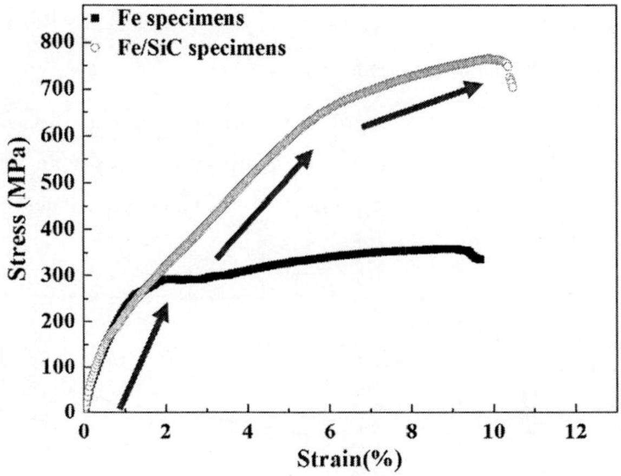

Figure 9.11: Stress–strain Curves of as-fabricated Fe/SiC Composite and as- fabricated Pure Fe Specimen [30]

In brief, the strengthening is a result of two major contributions, direct strengthening and indirect strengthening [33, 34]. Direct strengthening or classical composite strengthening, is based on load transfer from the matrix to the higher stiffness reinforcement particles (SiC). Indirect strengthening results from the changes in the matrix microstructure due to the presence of reinforcement particles.

Attar, H., et al [26] investigated the Selective laser melting of in situ titanium–titanium boride composites, Processing, and mechanical properties as discussed in section 9.7.1 Compressive mechanical testing was carried out on cylindrical parts along the SLM building direction (Z) using an Instron 5869 machine at a fixed cross-head speed of 0.001 mm/s following the specifications described in the standard DIN 50106. In general, a compressive test is not as susceptible as a tensile test to minor processing flaws which lead to premature failure. Therefore, compressive testing results can exhibit the intrinsic behavior (strength/work hardening) of the material and can be used to predict tensile properties [35].

Figure 9.12 shows the compressive true stress- strain curves of the Ti–TiB composite and CP-Ti processed by SLM. Initially, the flow stress of the Ti–TiB composite increases rapidly, reaching a maximum after exhibiting distinct strain hardening. Then, the flow stress drops with increasing strain until final failure.

A drop in stress may be attributed to the commencement of cracks. The true yield strength ($\sigma_{0.2}$) and the ultimate compressive strength (σ_{UCS}) of the Ti–TiB composite are significantly higher than the values for CP-Ti, indicating that the existence of 8.35 vol.% TiB greatly improves the compressive strength but not the plasticity as expected.

However, the material exhibits appreciable work hardening in comparison with the samples produced by conventional powder metallurgy technique. Moreover, as summarized in table 9.4, all compressive properties of the SLM-processed Ti–TiB composite are superior to those of the Ti–6Al–4V alloy [36].

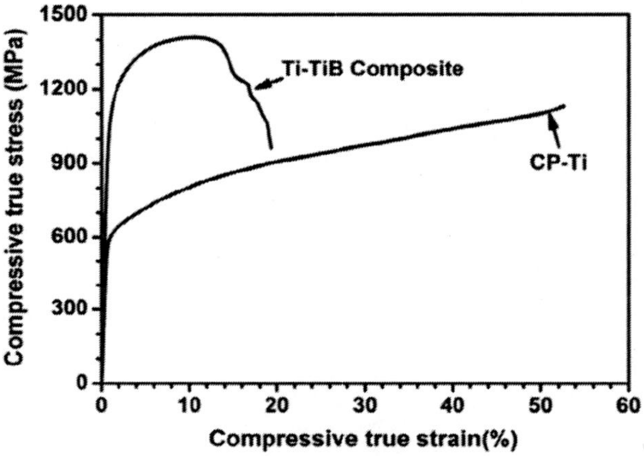

Figure 9.12: Compressive True Stress–strain Curves for CP-Ti and Ti–TiB Composite Produced by SLM [26]

Table 9.4: Compressive Properties of Biomedical Ti Materials [26]

Material type	**Processing method**	**Yield strength, $\sigma_{0.2}$ (MPa)**	**Ultimate Compressive strength, σ_{UCS} (MPa)**	**Maximum strain (%)**
Ti-8.35 vol.%TiB	SLM	1103±20	1421±47	1078±3.2
Ti-5 vol.%TiB	Powder metallurgy	1200	1420	8.5
Ti-10 vol.%TiB	Powder metallurgy	1400	1650	5.5
CP- Ti	SLM	560±5	1136±15	51±3.5
CP-Ti	SCAP	700	900	35
Ti-6Al-4V	Annealed	1000	1300	10

9.7. Tribological Behaviour of SLMed Parts

9.7.1. Wear

Gu, D et al [14] investigated the were performance of SLM processed TiC/TiSi3. The variations of friction coefficients and wear rates of SLM processed parts are provided in figure 9.13. The TiC/Ti5Si3 composite parts generally had much lower friction coefficients (the average value b0.5) as relative to the pure Ti part (the average value ~1.3). Also, the local fluctuation of the friction coefficients of the TiC/Ti5Si3 composites was considerably slighter than that of pure Ti part (figure 9.13a). A significantly improved wear performance of the TiC/Ti5Si3 parts was thus proved.

By comparison, the TiC/Ti5Si3 part prepared using 0.4 kJ/m demonstrated the superior wear performance. A uniform distribution of friction coefficient with a considerably low average value of 0.2 was obtained and the wear rate, 1.42×10^{-4} mm^3/Nm, was lowest as depicted in figure 6. This was essentially due to the homogeneous dispersion of ultrafine submicron- scale TiC dendrites and the sufficiently high densification level in this instance. Characteristic morphologies of the corresponding worn surfaces are revealed. At a relatively low η of 0.2 kJ/m, the worn surface of SLM-processed part primarily showed parallel grooves representing the abrasion wear. In this situation, the local severe plowing of the surface during sliding test resulted in a relatively high wear rate.

As η increased to 0.267 kJ/m, a majority of the worn surface was covered with adhesion tribolayer. On increasing η to 0.4 kJ/m, the worn surface of SLM-processed part became rather smooth. A continuously adherent, strain-hardened tribolayer without any significant fracture was formed on the worn surface.

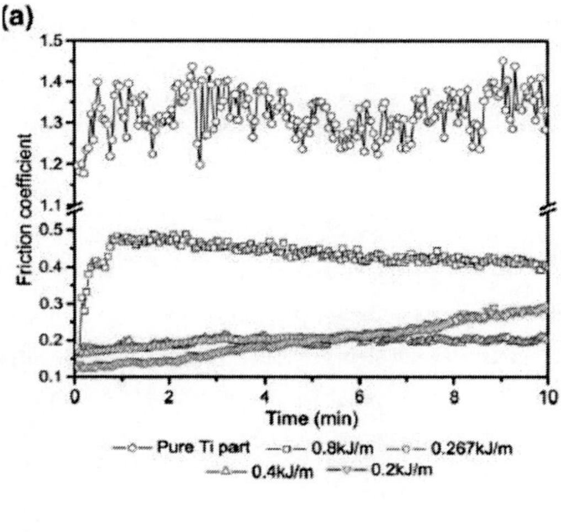

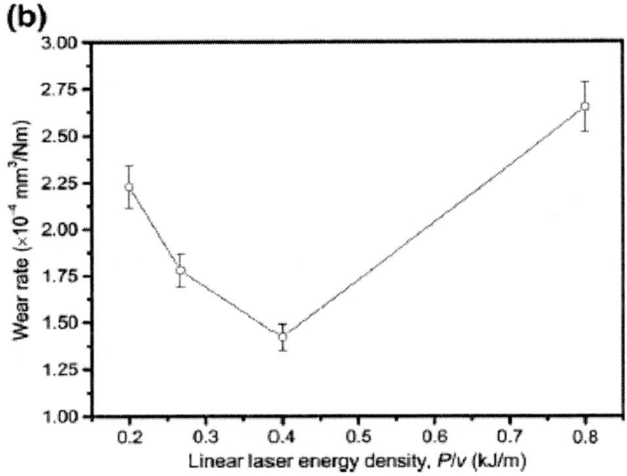

Figure 9.13(a) Variations of Friction Coefficient of Parts with different η(b) Wear Rate of Composite Parts with Applied Laser η [14]

Thus, it was reasonable to consider that as the applied laser energy density increased, the mechanism of material removal during sliding changed from the abrasion to the adhesion of tribolayer. Such a transition is expected to stabilize the friction coefficient and reduce the wear rate of SLM-processed parts, as demonstrated in figure 9.13. The worn surface of SLM-processed part at an even higher η of 0.8 kJ/m consisted of severely fragmented tribolayer and also entrapped debris. The wear rate in this case is nearly 2 times higher than that of the part processed at 0.4 kJ/m. A close look at the fine-grained and coarse-grained $TiC/Ti5Si_3$ composites at 0.4 and 0.8 kJ/m revealed that the significant grain coarsening of the TiC reinforcing phase at a higher laser energy density might account for the spalling of tribolayer and the resultant increase in wear rate.

9.7.2. Corrosion

Zhang, Y., et al [37] investigated the corrosion mechanism of amorphous alloy strengthened stainless steel composite by selective laser melting. Corrosion Science, 108241 with gas-atomized AISI 316L stainless steel and $Fe_{43.7}Co_{7.3}Cr_{14.7}Mo_{12.6}C_{15.5}B_{4.3}Y_{1.9}$ (at. %) amorphous alloy powders were used as the matrix and reinforcement, respectively. The chemical compositions of 316L SS and amorphous alloy powders are used. The 316L SS and amorphous alloy powders with the mass ratio of 90:10 ere blended, in which the particle size distribution of mixed powder was about 27–55 μm and the flowability of mixed powder was suitable for SLM processing. Using

EOS M1280 SLM machine which has a power of 400W (285 W used), focus diameter 70μm, 960mm/hatch spacing at 110 μm and the scanning direction in maintained at an angle of 67° between layers to fabricate a specimen of 10mm× 10mm× 5mm.

The cyclic potentiodynamic polarization (CPP) was conducted in the 3.5 wt. % NaCl solution at 25 °C. The specimens were reduced potentiostatically at −1.0 V for 3 min initially to remove the air-formed oxides on the surface before the polarization. Then, the specimens were stabilized in the electrolyte, where the open-circuit potentials (OCP) were measured. The CPP test was performed at 0.5 mV/s from-0.3 VAg/AgCl to a current density of 0.1 mA/cm^2, and then the scan direction was reversed. After that, the Ecorr and i_{corr} were calculated by,

Extrapolating the Tafel line. Each type of electrochemical measurement was repeated at least three times to ensure the reproducibility and consistency of the data. Later, the electrochemical impedance spectroscopy (EIS) measurements were performed at a potential of 0 V, and a perturbing sinusoidal potential was superimposed with amplitude of 10 mV. The anodic potential was approximately 0.6 V for the SLMed specimens, which is near the pitting potential (Epit). The difference of SLMed-composite and SLMed-SS can be demonstrated by the cyclic potentiodynamic polarization (CCP) measurement in 3.5% NaCl solution as shown in figure 9.14, in which a passive film can be formed and characterized in table 9.5.

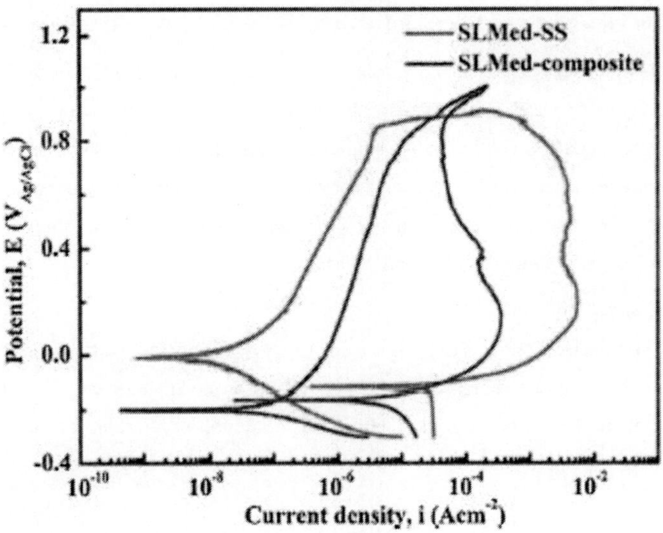

Figure 9.14: CPP Curves of SLMed-SS and SLMed-composite [37]

Table 9.5: Potentiodynamic Polarization Curve Parameter Values [37]

Specimens	E_{corr} (V)	E_{pit} (V)	E_{rep}(V)
As cast 316L	-0.132±0.006	0.591± 0.076	-0.125±0.015
SLMed SS	0 ±0.053	0.918 ± 0.035	-0.129±0.033
SLMed Composite	-0.172 ± 0.027	1.013 ± 0.010	-0.147 ± 0.018

We find that the SLMed-composite presents the widest passive region (1.185 $V_{Ag/AgCl}$) and the greatest pitting potential (E_{pit}) value (1.013 $V_{Ag/AgCl}$) over those of the SLMed-SS (0.918 $V_{Ag/AgCl}$, 0.918 $V_{Ag/AgCl}$) and 316L (0.723 $V_{Ag/AgCl}$, 0.591 $V_{Ag/AgCl}$). In addition, the repassivation potentials (E_{rep}) values of SLMed composite are greater than the self-corrosion potential (E_{corr}) values. Therefore, the passive film can be regenerated when the potential decreased even the passive film broke down at E_{pit}. However, the E_{rep} values of SLMed-SS are less than the E_{corr} values, demonstrating the formed passive film cannot be regenerated during the entire test once it was broken down at E_{pit}. This observation reveals that the pitting of SLMed-composite is easy to be suppressed compared to that of the SLMed-SS once the pitting occurred. Note that the reason of the higher E_{pit} for SLMed specimens can be ascribed to the homogenous microstructure by the rapid solidification rates in SLM (typically ~107 K/s). The fast solidification speed can decrease the diffusion time for Mn/S and then reduce the amount of formed MnS, thus inhibiting the pitting corrosion events initiated in the vicinity of second-phase particles [38, 39], as shown in table 9.5.

The anti-corrosion mechanism of the SLMed specimens was analyzed first by the EIS in figure 9.13. The Nyquist curves of the SLMed specimens show only one capacitive loop in the entire frequency range, which was fitted by the simplified equivalent circuit (inset of figure 9.13a). In the equivalent circuit, R_s and R_p represents solution resistance and charge-transfer resistance, respectively. CPE (constant-phase element) was used to represent the capacitance in the EIS analysis, and the impedance of CPE is defined as:

$$Z_{CPE} = [Y_0 (jw)^n]^{-1} \tag{9.5}$$

Where, Y_0 is the magnitude of CPE, j is the imaginary unit, ω is the angular frequency, n is the experience index. The value of n in SLMed- SS (0.90 ± 0.03) and SLMed-composite (0.86 ± 0.02) indicated that the CPE is close to capacitance, so the double layer capacitance (Cdl) can be calculated by the Brug's equation [40]:

$$C_{dl} = Y_0^{1/n}(R_s^{-1} + R_p^{-1})^{(n-1)/n} \tag{9.6}$$

The values of R_s, R_p, Y_0, n and C_{dl} were calculated by the ZView software, and the results are listed in table 9.5. R_p value relates to the passive film, indicating the corrosion resistance of the specimens. A lower R_p value means a higher electrochemical reaction rate. We find that the value of R_p in the SLMed-composite (4.38 ± 1.24MΩ cm^2) is greater than that of the SLMed-SS (1.35 ± 0.21MΩ cm^2). Besides, the value of C_{dl} in the SLMed-composite (2.15 ± 0.43 μF cm^{-2}) is smaller than that of the SLMed-SS (6.17 ± 2.27 μF cm^{-2}). These results demonstrate that a passive film was formed on the SLMed-composite and it has a greater corrosion resistance, which is consistent with the CPP measurements.

References

[1] Altenbach, H., Altenbach, J., & Kissing, W, *"Mechanics of Composite Structural Elements"*, 2004, ISBN 978-3-642-07411-0.

[2] http://compositeslab.com/composites-101/what-are-composites/#:~:text=A%20composite%20is%20a%20material,are%20a%20combination%20of%20components.

[3] http://home.iitk.ac.in/~mohite/Composite_introduction.pdf

[4] https://en.wikipedia.org/wiki/Composite_material#cite_note-73

[5] Wu, Xiangguo; Yang, Jing; Mpalla, Issa B, "Preliminary design and structural responses of typical hybrid wind tower made of ultrahigh performance cementitious composites", *Structural Engineering and Mechanics.* 48(6), 2013, pg. 791–807.

[6] Courtney, Thomas H, *"Mechanical Behavior of Materials"* Second Edition, 2002, Waveland Press. ISBN 978-1-4786-0838-7.

[7] *"Carbon Fiber Composite Design Guide"* (PDF), 2020.

[8] https://www.thoughtco.com/composites-in-aerospace-820418

[9] https://www.mech.utah.edu/composites_cars/

[10] https://compositesuk.co.uk/composite-materials/applications/automotive

[11] https://www.tuf-bar.com/composite-materials-for-civil-engineering-applications/

[12] Gourav Gupta, "Application and Future of Composite Materials: A Review" IJIRSET, vol. 5(5), 2016, pg. 6907- 6911.

[13] https://www.azom.com/article.aspx?ArticleID=8155

[14] Gu, D., Hagedorn, Y.C., Meiners, W., Wissenbach, K., & Poprawe, R, "Selective Laser Melting of in-situ TiC/Ti5Si3 composites with novel reinforcement architecture and elevated performance" Surface and Coatings Technology, 205(10), 2011, pg. 3285–3292.

[15] Zhou, X.B., & De Hosson, J.T.M, "Reactive wetting of liquid metals on ceramic substrates", *Acta Materialia,* 44(2), 1996, pg. 421–426.

[16] Zhao, X., Xu, S., & Liu, J, "Surface tension of liquid metal: role, mechanism and application", *Frontiers in Energy,* 11(4), 2017, pg. 535–567.

[17] Zhu, H., Lu, L., & Fuh, J.Y, "Influence of binder's liquid volume fraction on direct laser sintering of metallic powder", *Materials Science and Engineering: A,* 371(1-2), 2004, pg. 170–177.

[18] I. Takamichi, I.L.G. Roderick, *"The Physical Properties of Liquid Metals",* first ed. Clarendon Press, Oxford, 1993.

[19] Das, S, "Physical Aspects of Process Control in Selective Laser Sintering of Metals", *Advanced Engineering Materials,* 5(10), 2003, pg. 701–711.

[20] Dadbakhsh, S., & Hao, L, "Effect of Al alloys on selective laser melting behaviour and microstructure of in situ formed particle reinforced composites". *Journal of Alloys and Compounds,* 541, 2012, pg. 328–334.

[21] Dadbakhsh, S., Hao, L., & Sewell, N, "Effect of selective laser melting layout on the quality of stainless steel parts", *Rapid Prototyping Journal,* 18(3), 2012, pg. 241–249.

[22] Li, X.P., Kang, C.W., Huang, H., & Sercombe, T.B, " The role of a low-energy–density re-scan in fabricating crack-free Al85Ni5Y6Co2Fe2 bulk metallic glass composites via selective laser melting" *Materials & Design,* 63, 2014, pg. 407–411.

[23] Yan M, Wang JQ, Kong C, Schaffer GB, Qian M, "Micrometer-sized quasicrystals in the Al85Ni5Y6Co2Fe2 metallic glass: a TEM study and a brief discussion on the formability of quasicrystals in bulk and marginal glass-forming alloys", *J Mater Res;* 27, 2012, pg. 2131–2139.

[24] Mercelis P, Kruth J.P. Residual stresses in selective laser sintering and selective laser melting. *Rapid prototyp J;* 12, 2006, pg. 254–65.

[25] Mukai T, Nieh TG, Kawamura Y, Inoue A, Higashi K, "Dynamic response of a Pd40Ni40P20 bulk metallic glass in tension" *Scripta Mater;* 46, 2002, pg. 43–47.

[26] Attar, H., Bönisch, M., Calin, M., Zhang, L.C., Scudino, S., & Eckert, J, "Selective laser melting of in situ titanium–titanium boride composites: Processing, microstructure and mechanical properties", *Acta Materialia,* 76, 2014, pg. 13–22.

[27] Attar, H., Calin, M., Zhang, L.C., Scudino, S., & Eckert, J, "Manufacture by selective laser melting and mechanical behavior of commercially pure titanium", *Materials Science and Engineering: A,* 593, 2014, pg. 170–177.

[28] Morsi, K., & Patel, V.V, "Processing and properties of titanium–titanium boride (TiBw) matrix composites—a review", *Journal of Materials Science,* 42(6), 2007, pg. 2037–2047.

[29] Geetha, M., Singh, A.K., Asokamani, R., & Gogia, A.K, "Ti based biomaterials, the ultimate choice for orthopaedic implants – A review" *Progress in Materials Science,* 54(3), 2009, pg. 397–425.

[30] Song, B., Dong, S., Coddet, P., Zhou, G., Ouyang, S., Liao, H., & Coddet, C, "Microstructure and tensile behavior of hybrid nano-micro SiC reinforced iron matrix composites produced by selective laser melting", *Journal of Alloys and Compounds,* 579, 2013, pg. 415–421.

[31] Nardone, V.C., & Prewo, K.M, "On the strength of discontinuous silicon carbide reinforced aluminum composites", *Scripta Metallurgica,* 20(1), 1986, pg. 43–48.

[32] Miller, W., & Humphreys, F, "Strengthening mechanisms in particulate metal matrix composites", *Scripta Metallurgica et Materialia,* 25(1), 1991, pg. 33–38.

[33] Derby, B., & Walker, J, "The role of enhanced matrix dislocation density in strengthening metal matrix composites", *Scripta Metallurgica,* 22(4), 1988, pg. 529–532.

[34] Provenzano, V., Louat, N.P., Imam, M.A., & Sadananda, K, "On the validation of a strengthening concept" *Scripta Metallurgica et Materialia,* 24(11), 1990, pg. 2065–2070.

[35] Zhang, L.C., Das, J., Lu, H.B., Duhamel, C., Calin, M., & Eckert, J, "High strength Ti–Fe–Sn ultrafine composites with large plasticity", *Scripta Materialia,* 57(2), 2007, pg. 101–104.

[36] Srinivasan, K., & Venugopal, P, "Compression Testing of Ti-6Al-4V in the Temperature Range of 303–873 K", Materials and Manufacturing Processes, 23(4), 2008, pg. 342–346.

[37] Zhang, Y., Song, B., Ming, J., Yan, Q., Wang, M., Cai, C., Shi, Y, "Corrosion mechanism of amorphous alloy strengthened stainless steel composite by selective laser melting". *Corrosion Science,* 2019, 108241.

[38] Q. Chao, V. Cruz, S. Thomas, N. Birbilis, P. Collins, A. Taylor, P.D. Hodgson, D. Fabijanic, "On the enhanced corrosion resistance of a selective laser melted austenitic stainless steel", *Scr. Mater.* 141, 2017, pg. 94–98.

[39] M.P. Ryan, D.E. Williams, R.J. Chater, B.M. Hutton, D.S. McPhail, "Metallurgy (communication arising): stainless-steel corrosion and MnS inclusions", *Nature* 424 (6947), 2003, pg. 389–390.

[40] G. Brug, A. Van Den Eeden, M. Sluyters-Rehbach, J. Sluyters, "The analysis of electrode impedances complicated by the presence of a constant phase elemen", *J. Electroanal. Chem. Interfacial Electrochem.* 176 (1-2), 1984, pg. 275–295.

Chapter X

Conclusion and Future Scope

CHAPTER X

10. Conclusion and Future Scope

10.1. Conclusion

This book analyses the use of the SLM technique to achieve the prerequisite features on metallic materials. SLM is distinguished from conventional methods by process related factors, powder related factors and design related factors. The available research studies on SLM successfully demonstrate the feasibility of fabricating features with a required resolution on different materials including polymers, ceramics, and metals. An effective strategy for SLM would be to integrate all the subsystems such as powder dispensing, collection, and powder sieving with a closed-loop feedback system which leads to sustainable use of materials. An understanding of the microstructure and mechanical properties of parts fabricated using SLM is made with materials mainly metals. In view of the growing arenas for metallic parts with fine features in various fields, including automobile, aerospace, prosthetics, surgical guides, to implants in biomedical, structural and production process.

10.2. Future Scope

The forthcoming directional expansion of SLM should be focused on two aspects which are on equipment-related and process-related factor. An arrangement should be intended to handle nano sized metal powder particles, which tend to mix easily. The major focus should be on developing an innovative powder-recoating system at the same time not compromising on the speed. Regarding process knowledge, more studies are required in order to understand the interaction between nano scale powder particles and the laser beam interactions. Technological innovations have enabled the use of AM for metals at lower costs and have brought this process into small businesses. Due to different capabilities of processes, it can create shapes and parts that are not feasible through traditional manufacturing. Therefore, design methods need to be adapted for AM. Specifically, new design rules and tools are needed to (1) utilize topology optimization directly for part design; (2) use lattice and cellular structures in part design, and (3) use multi-material and different material distributions in part design. Current methods of specifying and verifying product quality are not sufficient to specify and verify the shapes and parts that can be produced using AM. Therefore, product quality specification and verification methods should be adapted to include the capabilities of AM techniques. In this future paradigm, integrated computational materials engineering will be leveraged to develop new materials for these new processes in parallel to their development and for specific applications. Similarly, the fine line between structural and functional AM will be merged and new devices that support customized medicine and the Internet of things will begin to emerge.

Space products are currently expensive and require a decade or more to infuse new technologies. Additive manufacturing enables >50% cost and schedule reduction for most parts driving affordable space exploration. Additive manufacturing enables improved system designs and space mission architectures that drive further toward sustainable and affordable space exploration. In future we can expect to see single material processes gain acceptance and also the multi-material processes infused into space products. We can also expect to see space-based additive manufacturing in wide use. With robust industry training, qualification programs, and low-cost flight demonstrations, supporters can greatly accelerate cultural acceptance and develop a new paradigm of affordable and sustainable space exploration.

In spite of significant developments and improvements in AM technologies, materials innovation using AM appears to be still in its embryonic stage. Comprehensive processing, microstructure, and property correlations are required. Major hurdle in designed materials development using AM is non-availability of desired materials. Promising results have been reported in developing metal, ceramic, and polymer matrix composites via AM techniques utilizing existing and new feedstock materials, several materials-related challenges remain to be addressed. Even existing materials are not optimized or designed for AM technologies. Important requirement in the development of new materials for AM is preparation of feedstock materials with desired characteristics suitable for specific AM technique. Finally, as with challenges the rewards are also extremely high for the integration of nano materials and AM. As one of the most popular and fastest growing additive manufacturing technologies, the laser-based powder bed fusion process involves a large number of processing parameters, which makes optimizing the process difficult. Like other additive manufacturing processes, the laser-based powder bed fusion process also provides research opportunities and manufacturing capacities in various fields:

1. Quick alloy design and development
2. Development of meta-stable stages in course of solidification under extreme cooling surroundings.
3. Design, manufacturing and characterization of parts with unique structures and sophisticated geometry, including cellular structures.

Further fundamental research into the SLM process, its numerical modelling analyses and in place of quality monitoring are still required to form this technology more robust and reliable for the manufacture of parts for structural applications.